Streamflow Depletion by Wells—Understanding and Managing the Effects of Groundwater Pumping on Streamflow

By Paul M. Barlow and Stanley A. Leake

Wyoming Pond on the Wood River, Pawcatuck River Basin, Rhode Island.

Photograph by Robert F. Breault, U.S. Geological Survey

Groundwater Resources Program

Circular 1376

U.S. Department of the Interior
U.S. Geological Survey

U.S. Department of the Interior
KEN SALAZAR, Secretary

U.S. Geological Survey
Marcia K. McNutt, Director

U.S. Geological Survey, Reston, Virginia: 2012

For more information on the USGS—the Federal source for science about the Earth, its natural and living resources, natural hazards, and the environment, visit *http://www.usgs.gov or call 1–888–ASK–USGS.*

For an overview of USGS information products, including maps, imagery, and publications, visit *http://www.usgs.gov/pubprod*

To order this and other USGS information products, visit *http://store.usgs.gov*

Suggested citation:
Barlow, P.M., and Leake, S.A., 2012, Streamflow depletion by wells—Understanding and managing the effects of groundwater pumping on streamflow: U.S. Geological Survey Circular 1376, 84 p.
(Also available at *http://pubs.usgs.gov/circ/1376/.*)

Library of Congress Cataloging-in-Publication Data

Barlow, Paul M.
  Streamflow depletion by wells: understanding and managing the effects of groundwater pumping on streamflow / by Paul M. Barlow and Stanley A. Leake.
    p. cm. -- (Circular ; 1376)
  Understanding and managing the effects of groundwater pumping on streamflow
  Includes bibliographical references.
  ISBN 978-1-4113-3443-4 (alk. paper)
  1. Streamflow. 2. Stream measurements. 3. Groundwater--Management. 4. Wells. I. Leake, S. A. II. Geological Survey (U.S.) III. Title. IV. Title: Understanding and managing the effects of groundwater pumping on streamflow. V. Series: U.S. Geological Survey circular ; 1376.
  GB1207.B373 2012
  628.1'12--dc23
                    2012038485

Foreword

Groundwater provides drinking water for millions of Americans and is the primary source of water to irrigate cropland in many of the Nation's most productive agricultural regions. Withdrawals in many aquifers throughout the United States have led to significant groundwater-level declines, resulting in growing concerns about sustainability and higher pumping costs. The U.S. Geological Survey's (USGS) Groundwater Resources Program has been instrumental in documenting groundwater declines and in developing groundwater-flow models for use in sustainably managing withdrawals.

Groundwater withdrawals also can lead to a reduction in streamflow, affecting both human uses and ecosystems. The first clear articulation of the effects of groundwater pumping on surface water was by the well-known USGS hydrologist C.V. Theis. In a paper published in 1940 entitled "The Source of Water Derived from Wells," Theis pointed out that pumped groundwater initially comes from reductions in aquifer storage. As pumping continues, the effects of groundwater withdrawals can spread to distant connected streams, lakes, and wetlands through decreased rates of discharge from the aquifer to these surface-water systems. In some settings, increased rates of aquifer recharge also occur in response to pumping, including recharge from the connected surface-water features. Associated with this decrease in groundwater discharge to surface waters is an increased rate of aquifer recharge. Pumping-induced increased inflow to and decreased outflow from an aquifer is now called "streamflow depletion" or "capture."

Groundwater discharge is a significant component of streamflow, with groundwater contributing as much as 90 percent of annual streamflow volume in some parts of the country. In order to effectively manage the entire water resource for multiple competing uses, hydrologists and resource managers must understand the effects (magnitude, timing, and locations) of groundwater pumping on rivers, streams, springs, wetlands, and groundwater-dependent vegetation.

This circular, developed as part of the USGS Groundwater Resources Program, presents concepts relating to streamflow depletion, methods for quantifying depletion, and common misconceptions regarding depletion. Approaches for monitoring, understanding, and managing streamflow depletion also are described. The report is written for a wide audience interested in the development, management, and protection of the Nation's water resources.

The Groundwater Resources Program provides objective scientific information and develops the interdisciplinary understanding necessary to assess and quantify the availability of the Nation's groundwater resources. Detailed assessments of regional aquifers have been completed in seven of the Nation's major aquifers, with several additional assessments ongoing or planned. The research and understanding developed through this program for issues such as streamflow depletion can provide the Nation's water-resource managers with the tools and information needed to manage this important natural resource.

Jerad D. Bales
Associate Director for Water (Acting)
U.S. Geological Survey

Contents

Boxes

Facing page: Rio Grande near Velarde, New Mexico. (Photograph by Michael Collier)

Conversion Factors and Datum

Inch/Pound to SI

Multiply	By	To obtain
Length		
foot (ft)	0.3048	meter (m)
mile (mi)	1.609	kilometer (km)
Area		
acre	4,047	square meter (m^2)
square foot (ft^2)	0.09290	square meter (m^2)
square inch (in^2)	6.452	square centimeter (cm^2)
square mile (mi^2)	2.590	square kilometer (km^2)
Volume		
acre-foot (acre-ft)	1,233	cubic meter (m^3)
cubic foot (ft^3)	0.02832	cubic meter (m^3)
gallon (gal)	3.785	liter (L)
million gallons (Mgal)	3,785	cubic meter (m^3)
Flow rate		
acre-foot per year (acre-ft/yr)	0.001233	cubic hectometer per year (hm^3/yr)
cubic foot per second (ft^3/s)	0.02832	cubic meter per second (m^3/s)
cubic foot per day (ft^3/d)	0.02832	cubic meter per day (m^3/d)
cubic foot per second per mile [(ft^3/s)/mi]	0.01760	cubic meter per second per kilometer [(m^3/s)/km]
cubic foot per second per square mile [(ft^3/s)/mi^2]	0.01093	cubic meter per second per square kilometer [(m^3/s)/km^2]
foot per second (ft/s)	0.3048	meter per second (m/s)
foot per day (ft/d)	0.3048	meter per day (m/d)
foot per day per foot [(ft/d)/ft]	1	meter per day per meter [(m/d)/m]
foot squared per day (ft^2/d)	0.09290	meter squared per day (m^2/d)
inch per year (in/yr)	25.4	millimeter per year (mm/yr)
gallon per minute (gal/min)	0.06309	liter per second (L/s)
million gallons per day (Mgal/d)	0.04381	cubic meter per second (m^3/s)

The following additional conversions for acre-foot per year (acre-ft/yr) are based on 365.25 days per year:

1 cubic foot per second (ft^3/s) is equal to 724.5 acre-ft/yr

1 million gallons per day (Mgal/d) is equal to 1,121.0 acre-ft/yr

Temperature in degrees Celsius (°C) may be converted to degrees Fahrenheit (°F) as follows: °F=(1.8×°C)+32

Temperature in degrees Fahrenheit (°F) may be converted to degrees Celsius (°C) as follows: °C=(°F−32)/1.8

*Transmissivity: The standard unit for transmissivity is cubic foot per day per square foot times foot of aquifer thickness [(ft^3/d)/ft^2]ft. In this report, the mathematically reduced form, foot squared per day (ft^2/d), is used for convenience.

Altitude, as used in this report, refers to distance above the vertical datum.

Streamflow Depletion by Wells—Understanding and Managing the Effects of Groundwater Pumping on Streamflow

By Paul M. Barlow and Stanley A. Leake

Introduction

Groundwater is an important source of water for many human needs, including public supply, agriculture, and industry. With the development of any natural resource, however, adverse consequences may be associated with its use. One of the primary concerns related to the development of groundwater resources is the effect of groundwater pumping on streamflow. Groundwater and surface-water systems are connected, and groundwater discharge is often a substantial component of the total flow of a stream. Groundwater pumping reduces the amount of groundwater that flows to streams and, in some cases, can draw streamflow into the underlying groundwater system. Streamflow reductions (or depletions) caused by pumping have become an important water-resource management issue because of the negative impacts that reduced flows can have on aquatic ecosystems, the availability of surface water, and the quality and aesthetic value of streams and rivers.

Scientific research over the past seven decades has made important contributions to the basic understanding of the processes and factors that affect streamflow depletion by wells. Moreover, advances in methods for simulating groundwater systems with computer models provide powerful tools for estimating the rates, locations, and timing of streamflow depletion in response to groundwater pumping and for evaluating alternative approaches for managing streamflow depletion. The primary objective of this report is to summarize these scientific insights and to describe the various field methods and modeling approaches that can be used to understand and manage streamflow depletion. A secondary objective is to highlight several misconceptions concerning streamflow depletion and to explain why these misconceptions are incorrect.

Lower Colorado River and adjacent farmland in the Yuma, Arizona, area. Diversion structure in upper right is Morelos Dam, the main point of delivery of water to Mexico. The "Law of the River" recognizes that water can be withdrawn from the Colorado River by "underground pumping." (Photograph by Andy Pernick, Bureau of Reclamation)

Characteristics of Groundwater Systems and Groundwater Interactions with Streamflow

This section provides brief descriptions of several terms and concepts that contribute to an understanding of streamflow depletion by wells. For a more extensive discussion of these concepts, the reader is referred to texts on groundwater, hydrogeology, and hydrology by Freeze and Cherry (1979), Linsley and others (1982), Heath (1983), Domenico and Schwartz (1990), and Fetter (2001).

Aquifers and Groundwater Flow

The pores, fractures, and other voids that are present in the sediments and rocks that lie close to the Earth's surface are partially to completely filled with water. In most locations, an unsaturated zone in which both water and air fill the voids exists immediately beneath the land surface (fig. 1). At greater depths, the voids become fully saturated with water. The top of the saturated zone is referred to as the water table, and the water within the saturated zone is groundwater.

Although voids beneath the water table are filled with water, the ability of subsurface materials to store and transmit water varies substantially. The term aquifer refers to subsurface deposits and geologic formations that are capable of yielding usable quantities of water to a well or spring, whereas a confining layer (or confining bed, such as illustrated in figure 1) refers to a low-permeability deposit or geologic formation that restricts the movement of groundwater (Heath, 1983). An aquifer can refer to a single geologic layer (or unit),

a complete geologic formation, or groups of geologic formations (Freeze and Cherry, 1979).

Most aquifers are classified as either confined or unconfined. A confined aquifer is one that lies between two confining layers, whereas an unconfined aquifer is one in which the uppermost boundary is the water table (fig. 1). Unconfined aquifers are often referred to as water-table aquifers, and both terms are used interchangeably in this report. As illustrated in figure 1, unconfined aquifers typically are located near land surface and confined aquifers are located at depth. Because of their proximity to land surface and associated surface waters, unconfined aquifers are often of interest in problems concerning streamflow depletion by wells; however, pumping from confined aquifers also can cause depletion. The fact that flow paths exist from deep confined aquifers upward to shallow aquifers means that changes in water levels from pumping (that is, drawdown) in deep confined aquifers also propagate to shallow aquifers with connected streams. An additional term, "leaky aquifer," is sometimes used to refer to an aquifer that receives inflow from adjacent low-permeability beds, although it is actually the adjacent beds that leak water to the aquifer (Freeze and Cherry, 1979).

In many areas of the United States, groundwater systems are composed of a vertical sequence of aquifers in which an upper, unconfined aquifer is underlain by a series of one or more confining beds and confined aquifers, such as is illustrated in figure 1. In many other areas, however, the groundwater system consists of a single, often unconfined, aquifer underlain by geologic formations, such as crystalline rock, whose permeabilities are so low that the formation can be assumed to be impermeable to groundwater flow. Aquifers of this type are used throughout the report to illustrate many of the factors that affect streamflow depletion by wells.

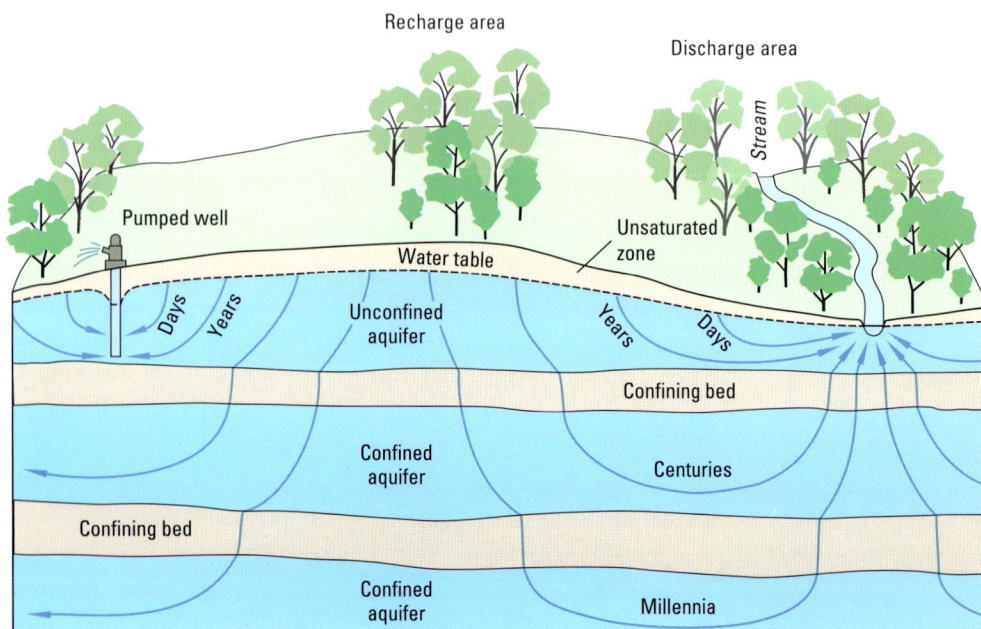

Figure 1. Groundwater flow paths in a multi-aquifer groundwater system. Groundwater flows from recharge areas at the water table to discharge locations at the stream and well. The residence time of groundwater can range from days to millennia (modified from Winter and others, 1998).

EXPLANATION

- - - - - - - - Water table
——— *90* ——— Line of equal hydraulic head, in feet
———▶ Direction of groundwater flow

Piezometer
┼ Water level
└◀ Open interval

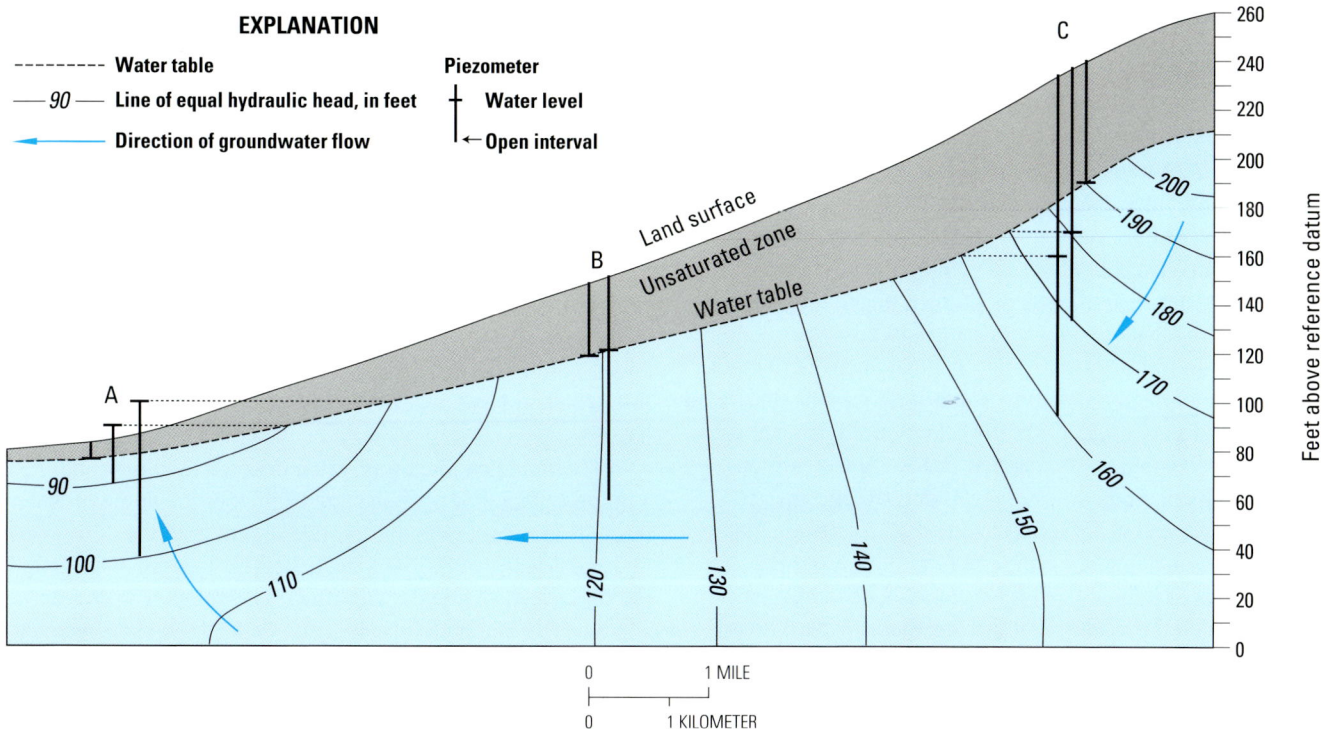

Figure 2. Distribution of hydraulic-head contours (groundwater levels) showing groundwater-flow directions in a vertical section of a hypothetical water-table aquifer. Groundwater levels are measured in piezometers, which are a type of observation well having a very short, open interval to the aquifer at the bottom of the well. The head measurements at the group of three piezometers completed at different depths at location C indicate downward groundwater flow at that location, whereas head measurements at the piezometers at locations B and A indicate lateral and upward flow at those locations, respectively (modified from Winter and others, 1998).

Groundwater moves continuously through aquifers from areas of groundwater recharge to areas of groundwater discharge. Such flow is illustrated by the flow paths in figure 1. The upper, unconfined aquifer shown in figure 1 is recharged by water that infiltrates across the land surface and then moves downward through the unsaturated zone to the water table to become groundwater. The source of groundwater recharge typically is precipitation (rain or melted snow) but can also originate from anthropogenic sources such as infiltration of irrigation return flow and septic-system wastewater. The accretion of water at the top of the saturated zone causes the water table to rise, and as a result, the saturated thickness of the unconfined aquifer increases. As recharge diminishes or ceases, the water table will decline and the saturated thickness decrease.

Groundwater commonly discharges to streams and wells, as illustrated in figure 1, but it can also discharge to springs, lakes, and ponds; to estuaries and directly to oceans; and by evaporation and plant transpiration in low-lying areas where the water table lies close to land surface, such as in wetlands or near streams. The residence time of water in a groundwater system can range from days to a few years for water recharged close to discharge boundaries, to millennia for water that travels along deep flow paths through low-permeability materials.

Directions of groundwater flow are determined from measurements of the altitude of groundwater levels made in wells. The water-level altitudes must be determined relative to a common datum plane, such as the National Geodetic Vertical Datum of 1929 (commonly referred to as "sea level;" Heath, 1983). Groundwater levels are equivalent to hydraulic heads and reflect the total potential energy of the groundwater system at the point of measurement. In a manner similar to flow in other potential fields (such as in electrical or thermal systems), groundwater flows from locations of higher potential energy to locations of lower potential energy and, therefore, in the direction of decreasing hydraulic head (fig. 2).

The rate of groundwater flow in a particular direction is dependent on the hydraulic conductivity of the aquifer, which is described in the next section, and the gradient of the hydraulic head in the direction of interest. The hydraulic gradient, which is equal to the change in head over a unit distance, can be determined from pairs of water-level measurements or from water-level contours drawn for a horizontal or vertical section of an aquifer. The hydraulic gradient between the 130 and 120 feet (ft) contours shown in figure 2, for example, is approximately 10 feet per mile, as determined by the change in hydraulic head between the two contours divided by the approximate distance between the contours along the flow line.

Groundwater systems are referred to as being in either a steady-state or a transient condition (fig. 3). A steady-state system is one in which groundwater levels and flow rates within and along the boundaries of the system are constant with time, and the rate of storage change within the flow system is zero. A transient system is one in which groundwater levels and flow rates change with time and are accompanied by changes in groundwater storage. Transient conditions occur in response to changes in flow rates along the boundaries of a groundwater system, such as short-term and long-term fluctuations in recharge rates, or changes in flow rates at points within a groundwater system, such as fluctuations in pumping rates. Although steady-state flow conditions, such as illustrated in figure 3A, rarely occur for real-world hydrologic conditions, it is often acceptable to assume that steady-state conditions exist if the fluctuations in water levels and storage changes are relatively small or if there is an interest in an evaluation of the long-term average condition of the flow system. Many studies of regional aquifer systems, for example, are conducted with the assumption that steady-state conditions occurred prior to large-scale groundwater development. During the predevelopment period, average rates of natural recharge and discharge to the aquifers are assumed to have

been in long-term balance. Another term that is sometimes used to refer to the state of a groundwater system is dynamic equilibrium (or steady-oscillatory; Maddock and Vionnet, 1998), in which water levels and flow rates are variable over a period of time (such as a year) but vary in a pattern that is the same from one period to the next (fig. 3C).

Hydraulic Properties of Aquifers, Confining Layers, and Streambed Sediments

The flow and storage of water in a groundwater system depend strongly on the hydraulic properties of the aquifers and confining layers that make up the system. These properties, which are summarized in table 1, also affect the timing, locations, and rates of streamflow depletion.

Hydraulic conductivity, often denoted by the symbol K, is a property that describes the rate of flow of a volume of water through a unit area of aquifer under a unit gradient of hydraulic head (Heath, 1983). The measurement units of K are length per time, such as feet per day (ft/d). The value of hydraulic conductivity at a particular location depends on the characteristics of the porous material, such as the size and arrangement of the pores and fractures, and the density and viscosity of the water within the porous material. Hydraulic-conductivity

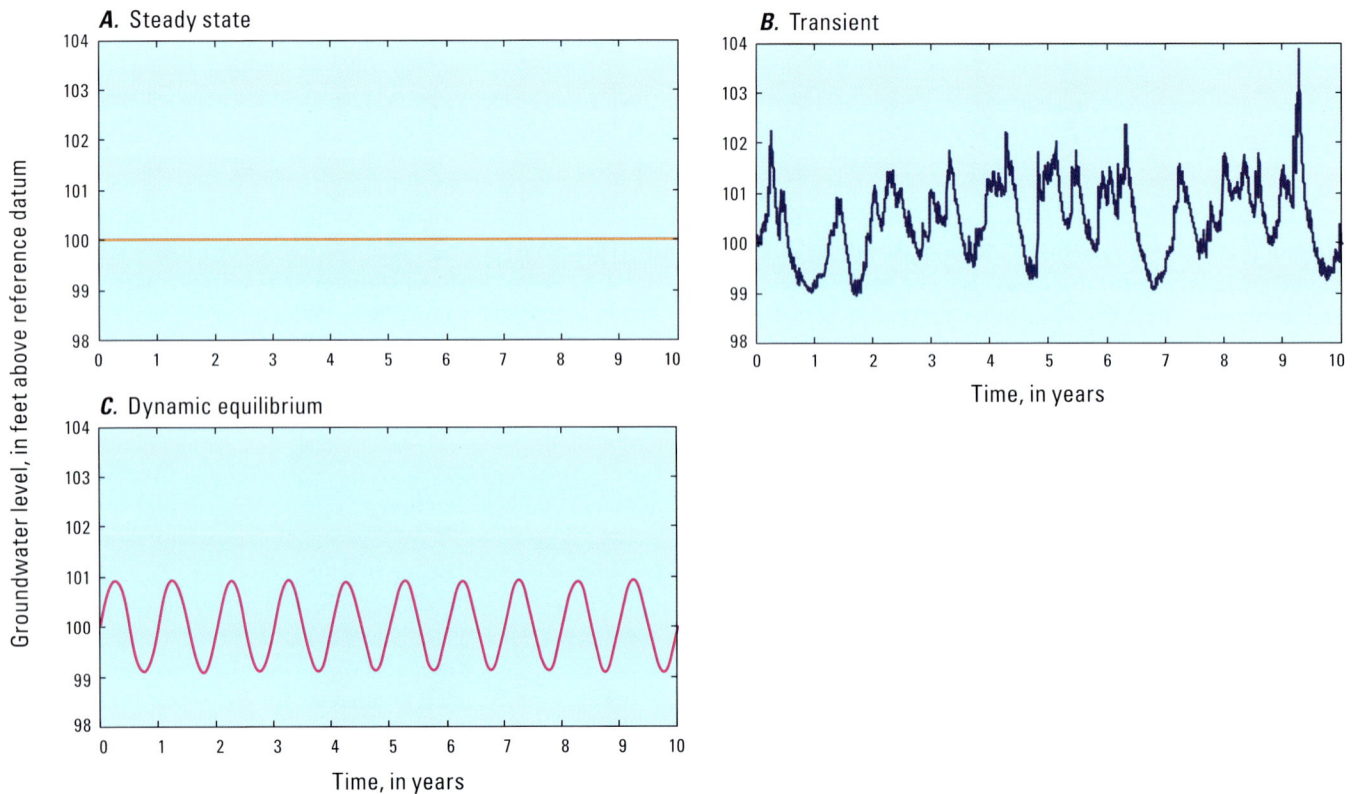

Figure 3.   Hydrologic conditions at a hypothetical observation well at which groundwater-level measurements indicate the state of the groundwater system. *A, Steady-state system:* Groundwater levels at the well do not change during the 10-year period. *B, Transient system:* Groundwater levels fluctuate with time, with the highest water levels generally occurring in the early spring and lowest water levels in the late summer and fall. *C, System in dynamic equilibrium:* Groundwater levels fluctuate throughout the year but in a pattern that is the same from one year to the next.

Table 1.   Aquifer properties that influence the timing of streamflow depletion.

Aquifer property	Symbol used	Units	Definition	Application
Saturated thickness	b	Length	The vertical thickness of the sediments in which pores are fully saturated	Any aquifer system
Hydraulic conductivity	K	Length/time	Rate of groundwater flow per unit area under a unit hydraulic gradient	Any aquifer system
Transmissivity	T	Length^2/time	Rate of groundwater flow per unit width under a unit hydraulic gradient $(T = K \times b)$	Aquifer systems dominated by horizontal flow
Specific storage	S_s	1/length	Volume of water released from or taken into storage per unit volume of aquifer per unit change in head	Confined aquifers
Storage coefficient	S	Dimensionless	Volume of water released from or taken into storage per unit surface area of aquifer per unit change in head normal to that surface $(S = S_s \times b)$	Confined aquifers
Specific yield	S_y	Dimensionless	Ratio of volume of water drainable by gravity from saturated aquifer material to the total volume of that material	Unconfined aquifers
Hydraulic diffusivity	D	Length^2/time	Ratio of the transmissivity to the storage properties of an aquifer: T/S, K/S_s, or T/S_y	Aquifer systems dominated by horizontal flow

values have a range of more than 12 to 13 orders of magnitude and are relatively large (~1 to 10,000 ft/d) for the unconsolidated sands and gravels and karstic limestones that typically constitute aquifers and relatively small (~1 × 10^{-8} to 0.1 ft/d) for clays, silts, and shales that typically constitute confining layers (Freeze and Cherry, 1979; Heath, 1983).

An aquifer in which the values of hydraulic conductivity differ from one location to another is said to be heterogeneous, whereas one in which the hydraulic conductivity is everywhere the same is said to be homogeneous. Although no natural aquifer is strictly homogeneous with respect to K, aquifer response to stress may in some cases be represented using a homogeneous equivalent K. For example, alluvial aquifers commonly include discontinuous beds of clay of low K distributed within sand of higher K. Even though the contrast in the hydraulic conductivity between the clay and sand may be orders of magnitude, the response to pumping may be approximated using a homogeneous K if the distribution of clay beds is uniform throughout the aquifer.

Unless specified differently, K refers to hydraulic conductivity in the horizontal direction. A more specific designation of horizontal hydraulic conductivity is K_h, and, similarly, vertical hydraulic conductivity commonly is designated as K_v. Because of the presence of low-permeability interbeds in many aquifers, K_h can be greater than K_v by a factor of 10 or more.

For groundwater systems that are dominated by horizontal flow, the transmissivity (T) at each location in an aquifer can be expressed as the product of the hydraulic conductivity and saturated thickness at that location: $T = K \times b$. Because the water table of an unconfined aquifer rises and falls in response to hydraulic stresses, such as recharge and pumping, the saturated thickness and transmissivity also vary in response to the changing water table. This complication is often of little consequence for thick aquifers where water-table fluctuations are relatively small, but may be important near pumping wells where water-table declines are a significant fraction of the initial saturated thickness. In such situations, and also in the case where pumping wells draw water from deep within an unconfined aquifer, vertical components of groundwater flow may be too large to ignore and the concept of transmissivity less useful.

The dominant process by which water is released from storage differs substantially between confined and unconfined systems (Heath, 1983). In confined aquifers and confining layers, water is released from storage by compression of the matrix of solid materials that form the deposit and by expansion of the water contained within the pores of the deposit. The storage capacities of confined aquifers and confining units are described by the hydraulic properties of specific storage (S_s) and storage coefficient (S), which are related by saturated thickness: $S = S_s \times b$. The storage properties of confined aquifers and confining units are relatively small; typical values of the storage coefficient of confined aquifers range from 5 × 10^{-5} to 5 × 10^{-3} (Freeze and Cherry, 1979). In contrast, the primary component of storage in an unconfined aquifer (or a confining layer that contains a water table) is drainage of water stored in the pores of the aquifer that is released as the water table declines. Water is also released from unconfined aquifers by compression of the aquifer matrix and expansion of the water, but these sources of stored water are small compared to drainage at the water table and typically are ignored. The storage capacity of an unconfined aquifer is described by its specific yield (S_y). The specific yields of unconfined aquifers are much larger than the storage coefficients of confined aquifers, typically between 0.01 and 0.30 (Freeze and Cherry, 1979).

Another hydraulic property that is not widely used in groundwater studies but has relevance to streamflow depletion is aquifer hydraulic diffusivity (D), which relates the transmissive and storage properties of an aquifer. Because of its importance to the timing and rates of streamflow depletion, it is described in detail in Box A.

Hydraulic properties of streambed and streambank materials may be different from those of the underlying aquifer or confining layer. The properties that are most important to the flow of water across the streambed and streambank materials are the hydraulic conductivity (K_s) and thickness (d_s) of the streambed sediments. In most analyses, the storage properties of these sediments are considered to be negligible.

Groundwater and Streamflow

Streams and rivers are commonly the primary locations of groundwater discharge, and groundwater discharge is often the primary component of streamflow. Groundwater is discharged through saturated streambed and streambank sediments, or permeable bedrock adjacent to the stream, where the altitude of the water table is greater than the altitude of the stream surface (fig. 4A). Conversely, streamflow seeps into the underlying groundwater system where the altitude of the stream surface is greater than the altitude of the adjoining water table (fig. 4B). Stream reaches that receive groundwater discharge are called gaining reaches and those that lose water to the underlying aquifer are called losing reaches. The rate at which water flows between a stream and adjoining aquifer depends on the hydraulic gradient between the two water bodies and also on the hydraulic conductivity of geologic materials that may be located at the groundwater/surface-water interface. A thick, silty streambed, for example, will tend to reduce the rate of flow between a stream and aquifer compared to a thin, sandy or gravelly streambed. In some cases, however, discharge from the aquifer to the stream is controlled by the rate at which groundwater must leave the aquifer. In this situation, the presence of a thick, silty streambed will tend to increase the hydraulic gradient between a stream and

A. Gaining stream

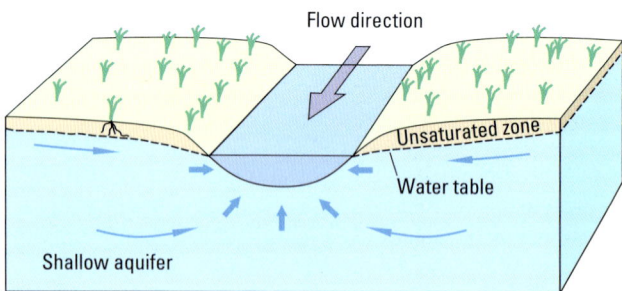

B. Losing stream

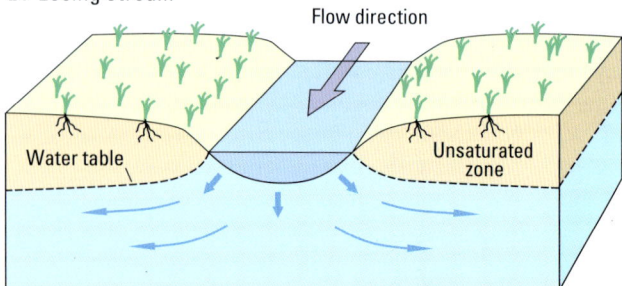

C. Gaining and losing reaches

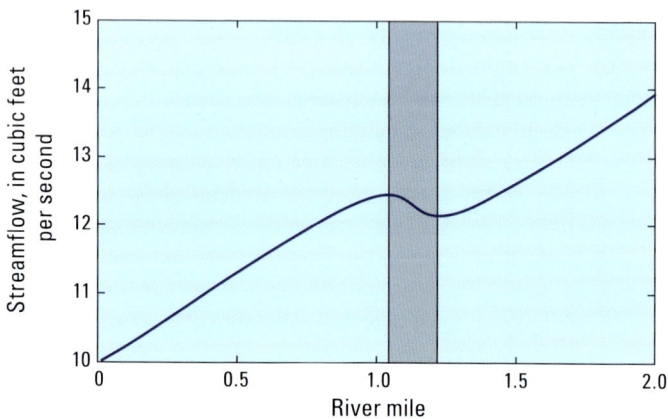

EXPLANATION

Gaining reach

Losing reach

Figure 4.   *A,* Gaining stream reaches receive water from the groundwater system, whereas, *B,* losing reaches lose water to the groundwater system. *C,* Streamflow increases along the gaining reaches of a river and streamflow decreases along the losing reaches of a river when there is no direct surface-water runoff to the river (parts *A* and *B* modified from Winter and others, 1998).

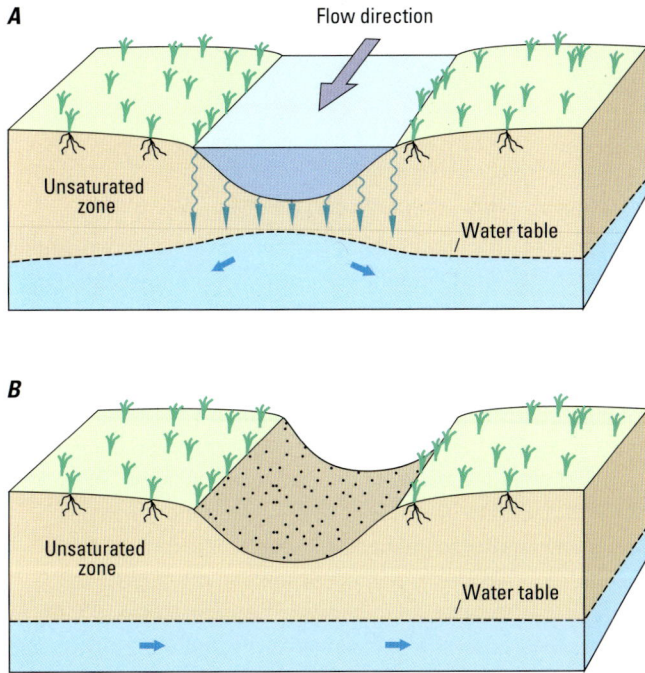

Figure 5. Disconnected stream reaches are separated from the groundwater system by an unsaturated zone. In *A*, streamflow is a source of recharge to the underlying groundwater system, but in *B*, streamflow and groundwater recharge have ceased (modified from Winter and others, 1998).

aquifer compared to the presence of a thin, sandy or gravelly bed, but will not affect the total amount of groundwater that is discharged to the stream.

The graph in figure 4*C* illustrates the effects of gaining and losing conditions on streamflow during a period of no direct surface-water runoff to the river. The graph shows that the rate of streamflow increases along gaining reaches and decreases along losing reaches. The graph also demonstrates that a stream can have both gaining and losing reaches simultaneously. Moreover, because precipitation rates, pumping rates, and other hydrologic stresses vary with time, it is possible for a particular stream reach to switch from a gaining to a losing condition or from a losing to a gaining condition from one period of time to the next.

Losing reaches occur under conditions in which the underlying sediments are fully saturated, as shown in figure 4*B*, or for conditions in which the sediments are unsaturated, as shown in figure 5*A*. A losing stream reach that is underlain by an unsaturated zone is said to be disconnected from the underlying aquifer (Winter and others, 1998). Some stream reaches are ephemeral (that is, they periodically become dry), and, as a consequence, flows between the stream and underlying aquifer may periodically cease (fig. 5*B*).

The sources of water to streams are generally recognized to result from four processes (Linsley and others, 1982): precipitation that falls directly onto a stream, which is a relatively small component of total streamflow; surface runoff (or overland flow) that travels over the land surface

Groundwater discharge from a basaltic-rock aquifer adjacent to the Metolius River, Deschutes River Basin, Oregon.

Box A: Hydraulic Diffusivity

Two of the most important factors that control the timing and rates of streamflow depletion are distance of the pumping well from connected surface waters and the hydraulic diffusivity of the aquifer. Distance to surface waters is easily understood, but hydraulic diffusivity is a less familiar property. Hydraulic diffusivity, D, is defined for confined aquifers as $D = T/S$, where T and S are the more familiar properties of transmissivity and storage coefficient, respectively.

The concept of aquifer diffusivity is strictly applicable to settings where water-level declines (drawdowns) from groundwater pumping propagate horizontally—but not vertically—to connected streams and other surface-water features. This condition implies that the saturated thickness of the aquifer remains constant over time, which is not the case for unconfined aquifers where the water table falls in response to pumping. Nevertheless, it is often acceptable to assume that changes in saturated thickness caused by pumping are relatively small (for example, less than 10 percent of the predevelopment saturated thickness) and that vertical groundwater-flow components within the aquifer are small compared to horizontal components. Under these assumptions, the hydraulic diffusivity of an unconfined aquifer is defined with respect to specific yield, S_y, as $D = T/S_y$.

Hydraulic stresses propagate faster through aquifers with higher values of hydraulic diffusivity than through aquifers with lower values of hydraulic diffusivity. It is important to understand that it is the ratio of T and S (or S_y) that controls the timing of depletion and not the values of T and S individually. For example, the rate of depletion at any given time caused by a pumping well in a system with a transmissivity of 10,000 feet squared per day (ft²/d) and a storage coefficient of 0.01 would be the same as in a system with a

transmissivity of 1,000 ft²/d and a storage coefficient of 0.001, assuming all other factors are equal. As illustrated in table A–1 for representative confined and unconfined aquifers with equal transmissivity, the hydraulic diffusivity of confined aquifers is typically several orders of magnitude greater than that for unconfined aquifers. This difference results from the much larger storage capacity of the unconfined aquifer (as represented by the value of specific yield) compared to that of the confined aquifer (as represented by the storage coefficient).

Higher values of hydraulic diffusivity increase the speed at which responses to stresses such as pumping propagate through an aquifer to connected streams. Streamflow depletion therefore generally will occur much more rapidly in confined aquifers than in unconfined aquifers (fig. A–1). Each of the responses shown in figure A–1 illustrates the slower and damped response to a pumping stress in an unconfined aquifer with a relatively low hydraulic diffusivity compared to the faster response to the same stress in a confined aquifer with a relatively high hydraulic diffusivity. The responses shown in the figure are characteristic of streamflow depletion from pumping, but hydraulic diffusivity similarly affects groundwater-level responses to stresses other than pumping, such as recharge and changes in surface-water stage.

A final point concerning the propagation of hydraulic stresses within an aquifer is that the rate of propagation of a hydraulic perturbation is not the same as the velocity with which a volume of groundwater actually travels through an aquifer or the associated residence time of groundwater in the aquifer. Groundwater movement is nearly always substantially slower than the propagation of hydraulic stresses through most types of aquifers, particularly those that are the source of most large-scale groundwater withdrawals.

Table A–1.   Example transmissivity, storage property, and resulting hydraulic diffusivity of a confined and unconfined aquifer.

[Saturated thickness *(b)* of both aquifers is 100 feet, hydraulic conductivity *(K)* is 100 feet per day, and specific storage *(S_s)* is 1×10^{-6} feet⁻¹; –, not applicable]

Aquifer type	Transmissivity *(K × b)* [feet squared per day]	Storage coefficient *(S_s × b)* [dimensionless]	Specific yield [dimensionless]	Hydraulic diffusivity [feet squared per day]
Confined	10,000	0.0001	–	1×10^8
Unconfined	10,000	–[1]	0.1	1×10^5

[1]Although storage changes related to the product of S_s and b apply in unconfined aquifers, this property can be ignored in analyses of responses to pumping if the product is much smaller than specific yield.

Stress to aquifer

Response some distance from the applied stress

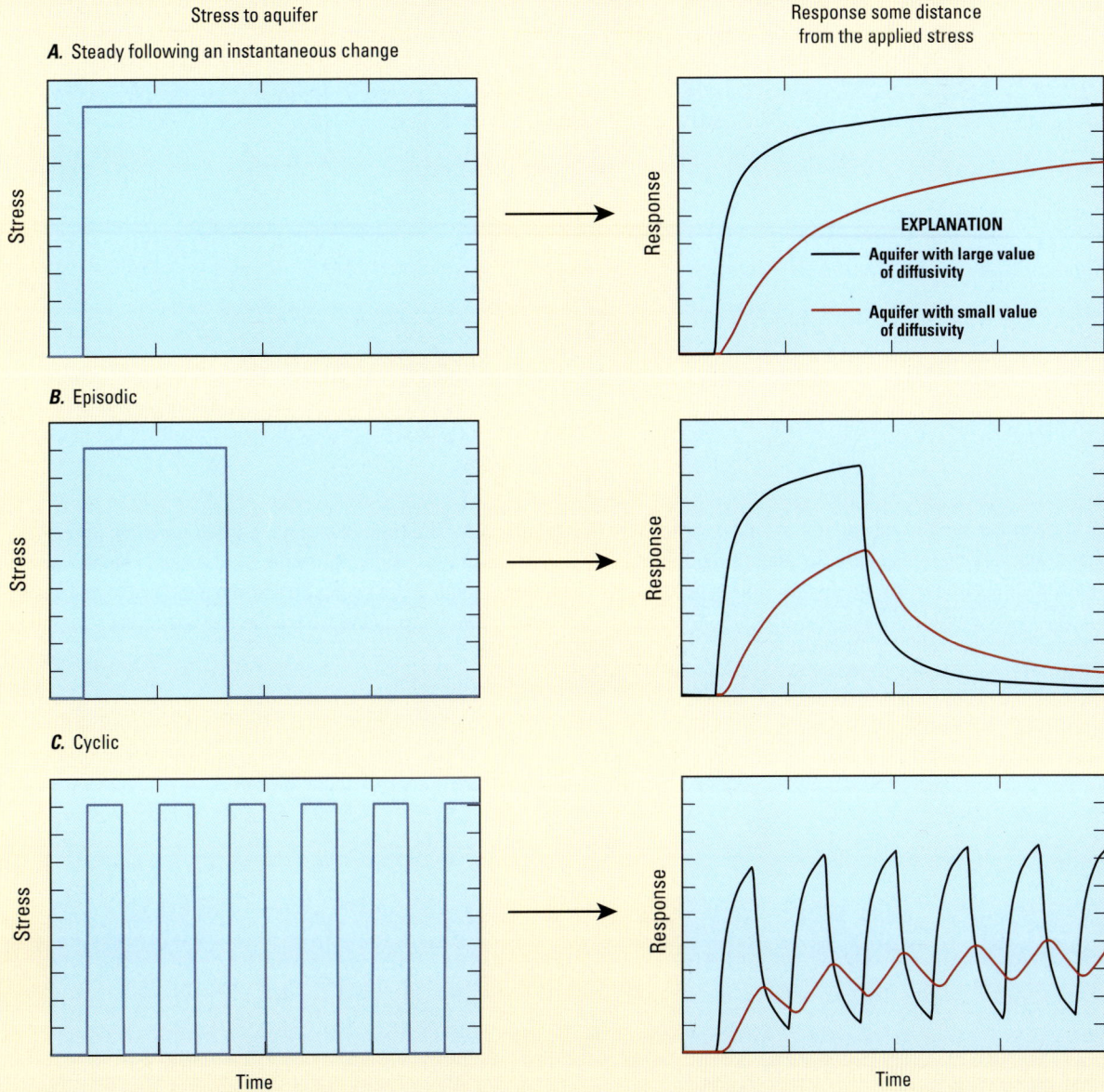

Figure A–1. Groundwater-system response to different types of stresses for two values of hydraulic diffusivity—a relatively large value representative of confined aquifers and a relatively small value representative of unconfined aquifers. Stresses to the aquifer could be pumping at a well or recharge to an aquifer. The responses shown are characteristic streamflow-depletion responses to pumping, but also could be water-level responses to pumping or recharge. For the hypothetical situation shown, the pumping stresses and streamflow-depletion responses would be in units of volume per time (such as cubic feet per second). Other types of stresses and resulting responses could have different measurement units; for example, recharge rates typically are reported in units of length per time (such as inches per year) and water levels in units of length (such as feet). [Rates of streamflow depletion were calculated by using a computer program described in Reeves (2008); hydraulic diffusivity of confined and unconfined aquifers are 100,000 and 10,000 feet squared per day, respectively.]

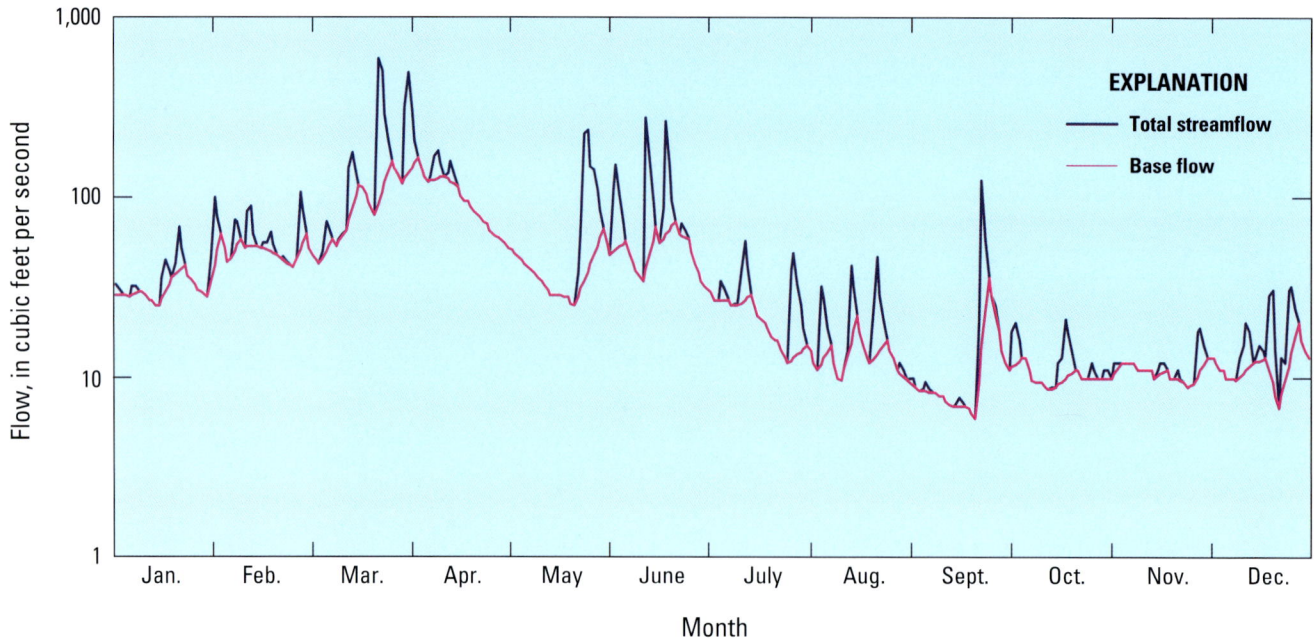

Figure 6.   Total streamflow and the estimated base-flow component of streamflow for the Hunt River near East Greenwich, Rhode Island, 2001. Temporally varying rates of precipitation, evaporation, and plant transpiration within the watershed result in highly variable rates of daily and seasonal streamflow conditions. During periods of streamflow decline, such as occurred from mid April to mid May, streamflow consists nearly entirely of base flow. The direct-runoff component of streamflow is the difference between total streamflow and base flow. (Data available from U.S. Geological Survey National Water Information System Web Interface, *http://waterdata.usgs. gov/nwis*; base flow estimated by the PART computer program documented in Rutledge, 1998.)

to a stream channel; interflow (or subsurface storm flow) that moves through the upper soil layers to a stream channel; and groundwater discharge, which is commonly referred to as base flow. Surface runoff and interflow are important during storm events, and their contributions typically are combined into a single term called the direct-runoff component of streamflow. Groundwater on the other hand is most important for sustaining the flow of a stream during periods between storms and during dry times of the year.

The proportion of streamflow that is contributed by groundwater discharge varies across physiographic and climatic settings (Winter and others, 1998). Base-flow contributions can be estimated for some streams by analysis of streamflow hydrographs, such as is illustrated for the Hunt

River in Rhode Island (fig. 6). During periods of streamflow decline (recession) that follow storms, streamflow in the river consists nearly entirely of groundwater discharge, but groundwater discharge also contributes to streamflow during and shortly after periods of high flow. The average long-term base-flow component of the Hunt River was estimated to be nearly 81 percent of the total flow in the river (Barlow and Dickerman, 2001). This large contribution of groundwater discharge is typical for rivers of the Northeastern United States that are underlain by highly permeable sand and gravel deposits that facilitate high rates of groundwater recharge and low rates of direct runoff. The contribution of groundwater discharge to streamflow is lower for basins underlain by less-permeable materials.

Streamflow Response to Groundwater Pumping

This section describes the fundamental processes and factors that affect the timing, rates, and locations of streamflow depletion. Unless otherwise stated, two important assumptions are made throughout this discussion—first, that the stream and underlying aquifer remain hydraulically connected by a continuous saturated zone, and second, that the stream does not become dry. These assumptions may not be valid for extreme cases of large-scale groundwater development and limited streamflow where groundwater levels have been drawn down below the bottom of the streambed. When the stream cannot supply the quantity of water pumped, the stream may eventually lose all of its water to the aquifer and become ephemeral. Even if flow remains in the stream, once groundwater levels decrease below the streambed, an unsaturated zone may develop near the locations of pumping that disconnects the groundwater and surface-water systems, at which time the flow rate between the groundwater and surface-water systems in the affected areas will no longer respond to pumping. Brunner and others (2011) provide a summary of several of the issues related to disconnected systems and the factors that influence the dynamics of disconnection; Su and others (2007) and Zhang and others (2011) provide examples of the effects of pumping on the formation of disconnected systems.

Time Response of Streamflow Depletion During Pumping

As stated by Theis (1940) in his seminal work on the source of water derived from wells, knowledge of the influence of time is fundamental to understanding the effects of groundwater development on aquifers and hydraulically connected surface waters. When a well begins to pump water from an aquifer, groundwater levels around the well decline, creating what is called a "cone of depression" in the water levels around the well. These water-level declines are largest at the well and decrease to effectively zero decline at some radial distance from the well (fig. 7). The hydraulic gradient that is established within the cone of depression forces water to move from the aquifer into the well. Initially, all of the water pumped by the well comes from water stored in the aquifer. The cone of depression generally deepens and expands laterally with increased pumping time. Because the hydraulic diffusivity of confined aquifers is relatively large, the cone of depression that forms around a well in a confined aquifer expands rapidly away from a well. In contrast, because the hydraulic diffusivity of unconfined aquifers is relatively small, the cone of depression around a well pumping from an unconfined aquifer expands slowly outward from the well.

The release of water from aquifer storage continues to be the only source of water to the well until the cone of depression reaches one or more areas of the aquifer from which water can be captured. Captured water consists of two possible sources—a reduction in the natural discharge (or outflow) rate of groundwater from the aquifer or an increase in the natural or artificial recharge (or inflow) rate to the aquifer. The primary sources of captured discharge are groundwater that would otherwise have flowed to streams, drains, lakes, or oceans, as well as reductions in groundwater evapotranspiration in low-lying areas such as riparian zones and wetlands. Figure 7C illustrates the capture of groundwater that would otherwise have discharged to the bounding stream. Groundwater discharge to the stream is reduced because groundwater levels at the stream-aquifer boundary have been lowered by pumping, which reduces the hydraulic gradient from the

Photograph by David E. Burt, Jr., U.S. Geological Survey

Groundwater pumped for flood irrigation of a rice field. Groundwater withdrawals from the Mississippi River alluvial aquifer to support agriculture in the Mississippi Delta region have resulted in groundwater-level declines and reductions in groundwater discharge to many Delta streams (Barlow and Clark, 2011).

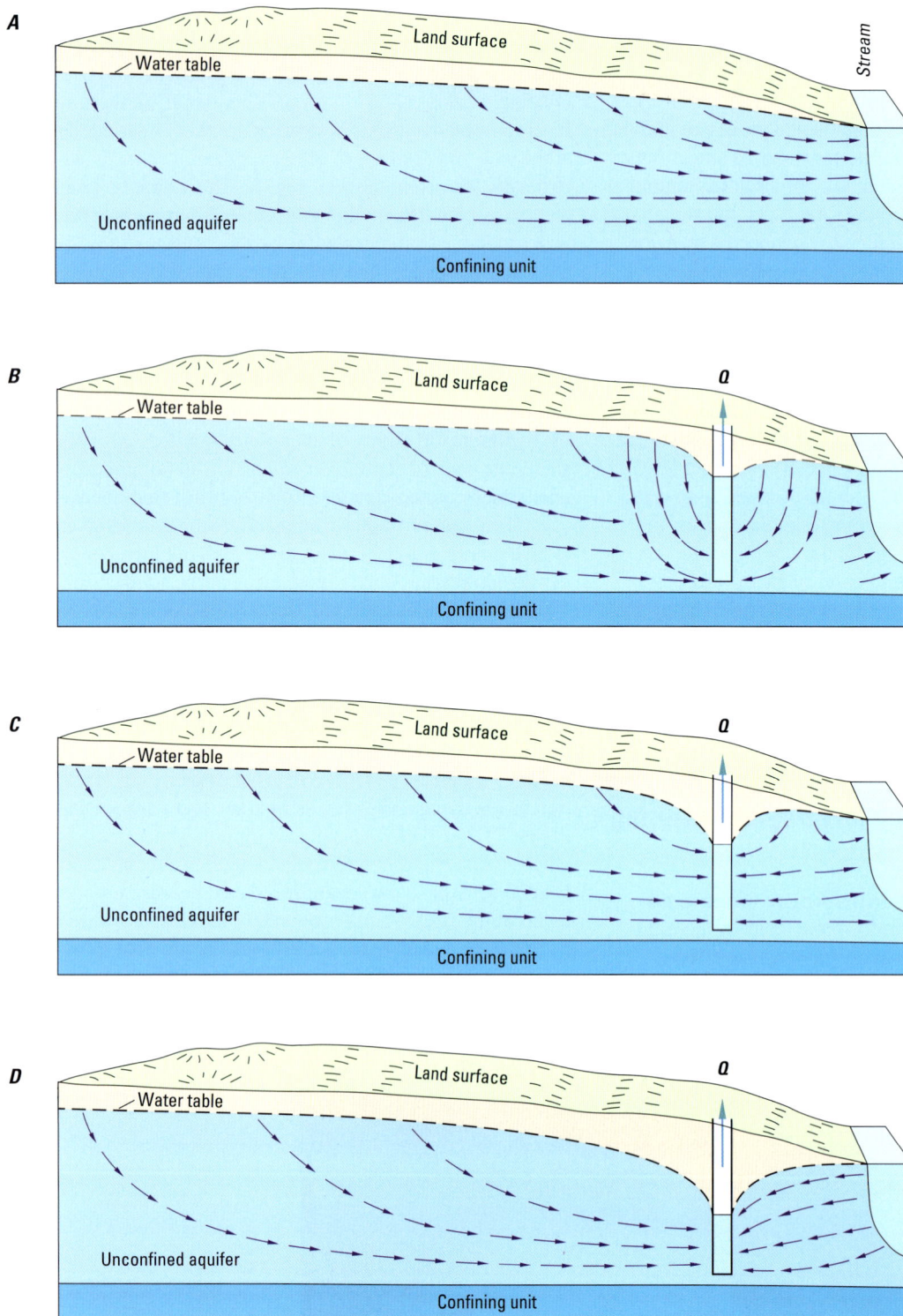

Figure 7. Effects of pumping from a hypothetical water-table aquifer that discharges to a stream. *A,* Under natural conditions, recharge at the water table is equal to discharge at the stream. *B,* Soon after pumping begins, all of the water pumped by the well is derived from water released from groundwater storage. *C,* As the cone of depression expands outward from the well, the well begins to capture groundwater that would otherwise have discharged to the stream. *D,* In some circumstances, the pumping rate of the well may be large enough to cause water to flow from the stream to the aquifer, a process called induced infiltration of streamflow. Streamflow depletion is equal to the sum of captured groundwater discharge and induced infiltration (modified from Heath, 1983; Alley and others, 1999). [*Q,* pumping rate at well]

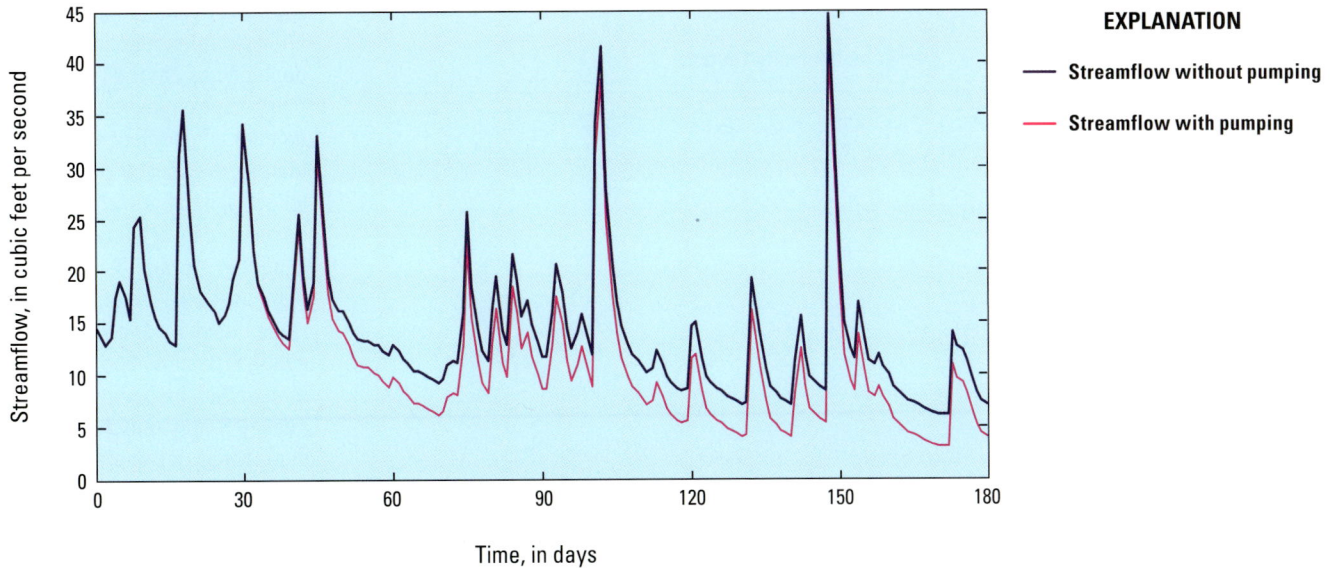

Figure 8. Effects of groundwater pumping on a hypothetical streamflow hydrograph. Top curve shows daily streamflow without pumping at a nearby well. Lower curve shows daily streamflow with pumping from a well located near the stream at a rate of 2.0 million gallons per day (about 3.1 cubic feet per second) beginning at day 30. After about day 60, the total decrease in streamflow each day is equal to the pumping rate of the well.

aquifer to the stream; however, there is no reversal in the gradient toward the stream, and the stream remains gaining. An example of captured recharge is induced leakage from streams, drains, or lakes. For example, if the reductions in groundwater levels near a hydraulically connected stream are large enough, the hydraulic gradient at the stream-aquifer interface will be reversed, and streamflow will be induced to flow into the aquifer toward the well (fig. 7D). This process is referred to as induced infiltration of streamflow and results in the stream becoming losing within the reach of stream in which the gradient has been reversed. Captured groundwater discharge to streams and induced infiltration of streamflow both result in reductions in the total amount of streamflow; as a result, the two processes are combined into the single term streamflow depletion. Reductions in streamflow that result from pumping at a hypothetical well are illustrated for a representative streamflow hydrograph in figure 8. The lower curve on the graph illustrates that streamflow continues to rise and fall in response to precipitation events, but the rates of streamflow are lower than those that would occur in the absence of pumping. For the hypothetical conditions shown, the amount of streamflow reduction at any point in time is equal to the pumping rate of the well after about 60 days of pumping.

The time response of the sources of water to a hypothetical well is illustrated by the curves in figure 9. For this example aquifer, the only sources of water to the well are groundwater released from aquifer storage and streamflow depletion in a nearby stream. Groundwater storage is the primary source of water to the well soon after pumping begins, but its contribution to the well's withdrawal declines with time. The time at which more than half of the pumping

rate of the well is supplied by streamflow depletion is designated on the figure as the time to reach a depletion-dominated supply (t_{dds}). If the well pumps for an extended period of time, the source of water pumped by the well will be entirely from depletion, with no further contributions from groundwater storage. When this occurs, water levels no longer decline in response to pumping, the cone of depression does not expand any further, and the aquifer is in a new state of equilibrium in which the pumping rate of the well is equal to the amount of streamflow depletion. The time that is required for a new state of equilibrium to be attained has been called the "time to full capture" and can range from a matter of days to decades and even centuries (Bredehoeft and Durbin, 2009; Walton, 2010). In some aquifers, however, a new equilibrium may never be reached if the total pumping rate from the aquifer exceeds the rate at which water can be captured. In other aquifers, the time to reach full capture, as expressed as 100 percent of the pumping rate of the well, is so long that for practical purposes it is not meaningful. In those cases, it may be preferable to define full capture as a value somewhat less than 100 percent, such as 99 percent or 95 percent.

The factors that control the time response of streamflow depletion to pumping are the geologic structure, dimensions, and hydraulic properties of the groundwater system; the locations and hydrologic conditions along the boundaries of the groundwater system, including the streams; and the horizontal and vertical distances of wells from the streams. The effects of these factors will be illustrated in different ways throughout this report, beginning with a discussion of two of the most important variables—the distance of a pumping well from a nearby stream and the hydraulic diffusivity of an aquifer.

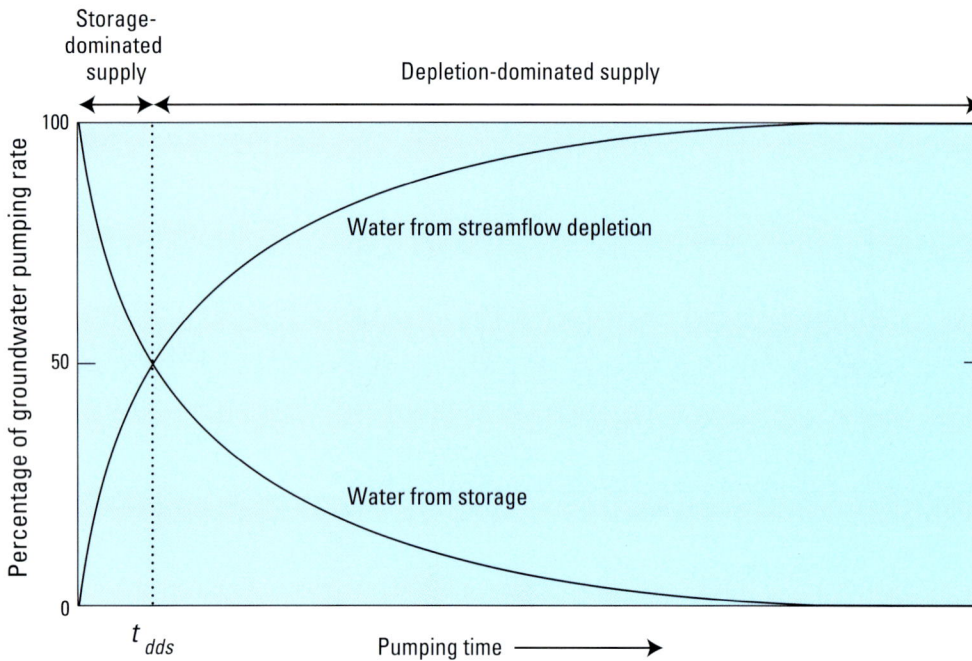

Figure 9. Relation of storage change and streamflow depletion as sources of pumped groundwater through time for a hypothetical well. Initially, the source of water (or supply) to the well is dominated by reductions in aquifer storage. At later times, streamflow depletion is the dominant source of supply. The condition of more than half of the pumping rate coming from streamflow depletion is designated as depletion-dominated supply, and variable t_{dds} is the time to reach the condition of depletion-dominated supply for a particular pumping location.

Jenkins (1968a, b) introduced a term that is widely applied in streamflow-depletion problems called the "stream depletion factor" (or SDF) to quantify the relation between these two variables. The stream depletion factor for a well pumping at a particular location in an aquifer is defined as

$$SDF = \frac{d^2}{D},$$

where d is the shortest distance between the pumped well and nearby stream, and D is the hydraulic diffusivity of the aquifer. Values of SDF have units of time.

For a given pumping location, the value of SDF is a relative measure of how rapidly streamflow depletion occurs in response to a new pumping stress. Streamflow depletion will occur relatively quickly in response to pumping from wells with a low value of SDF and relatively slowly in response to wells with a high value of SDF. A high value of hydraulic diffusivity, for example, will result in a relatively low value of SDF and, as described and illustrated in Box A, a relatively fast response of streamflow depletion to pumping. The effects of well distance on streamflow depletion are illustrated in figure 10 for two hypothetical wells pumping from the same aquifer. Because well A is located much farther from the stream than well B, the time necessary for the cone of depression formed by pumping at well A to reach the stream is much longer than that for well B, and as a result, groundwater-storage depletion is a source of water to the well for a longer period of time. In contrast, the cone of depression formed by pumping at well B reaches the stream much sooner than that

for well A, and streamflow depletion becomes the primary source of water to the well much sooner than for well A.

The presence of streambed and streambank sediments that impede the flow of water at the stream-aquifer interface also can affect the response of streamflow to pumping (fig. 11). These bed sediments often consist of fine-grained deposits and organic materials that have a lower hydraulic conductivity (permeability) than the surrounding aquifer materials. The effect of these sediments is to extend the time to full capture and to reduce the amount of streamflow depletion that occurs at any given time relative to a condition in which the low-permeability sediments are absent. For example, for the simulated conditions shown in the graph in figure 11, 65 percent of the water withdrawn by the well after 60 days of pumping consists of streamflow depletion for the condition with no resistance to flow at the stream-aquifer interface, whereas only 53 percent of the well's withdrawal rate consists of streamflow depletion after 60 days of pumping for the condition in which streambed and streambank materials with lower permeability than the aquifer are present at the stream-aquifer interface.

Conditions that do not affect the timing of depletion also are worth noting. First, in most aquifer systems, the timing of streamflow depletion is independent of the pumping rate at the well. If the pumping does not cause system changes such as large reductions in aquifer thickness or the drying up of streams or wetlands, depletion at any given time is proportional to the pumping rate. Depletion, therefore, can be expressed as a fraction (or percentage) of the pumping rate at a well, as described in Box B. Moreover, the fraction

A

B

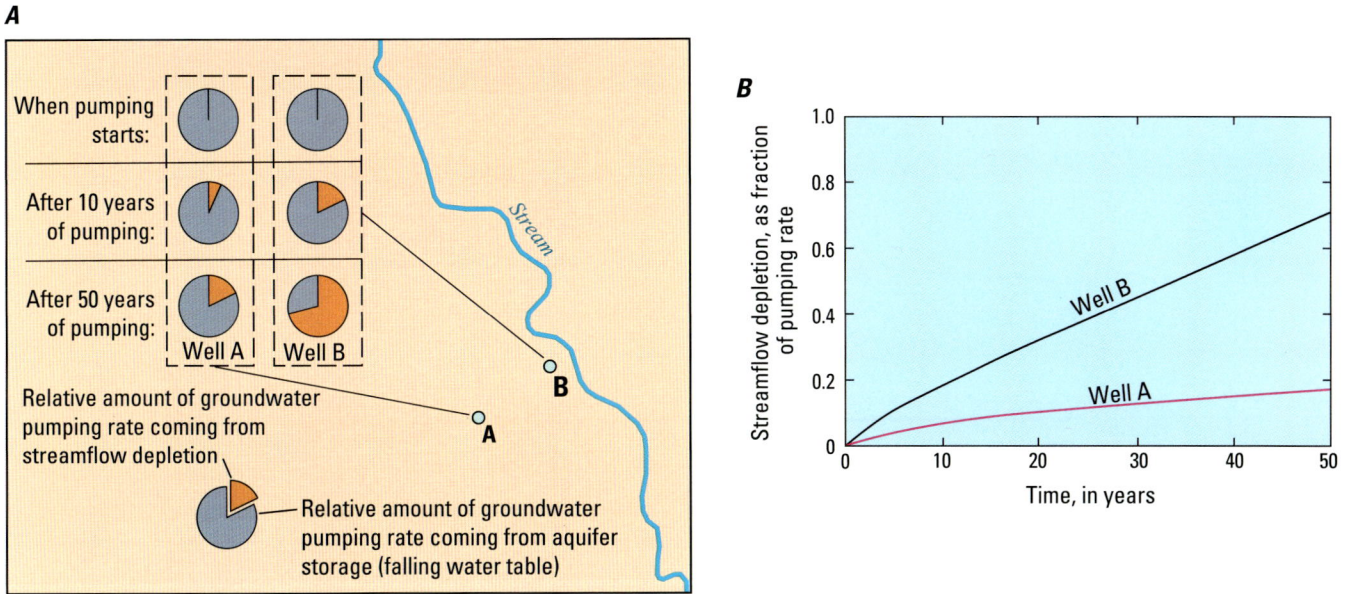

Figure 10. *A,* Sources of pumped groundwater at two hypothetical well locations for pumping times of 10 and 50 years. *B,* Streamflow depletion is a much larger source of water to well B than to well A during the 50-year pumping period because well B is much closer to the stream (modified from Leake and Haney, 2010).

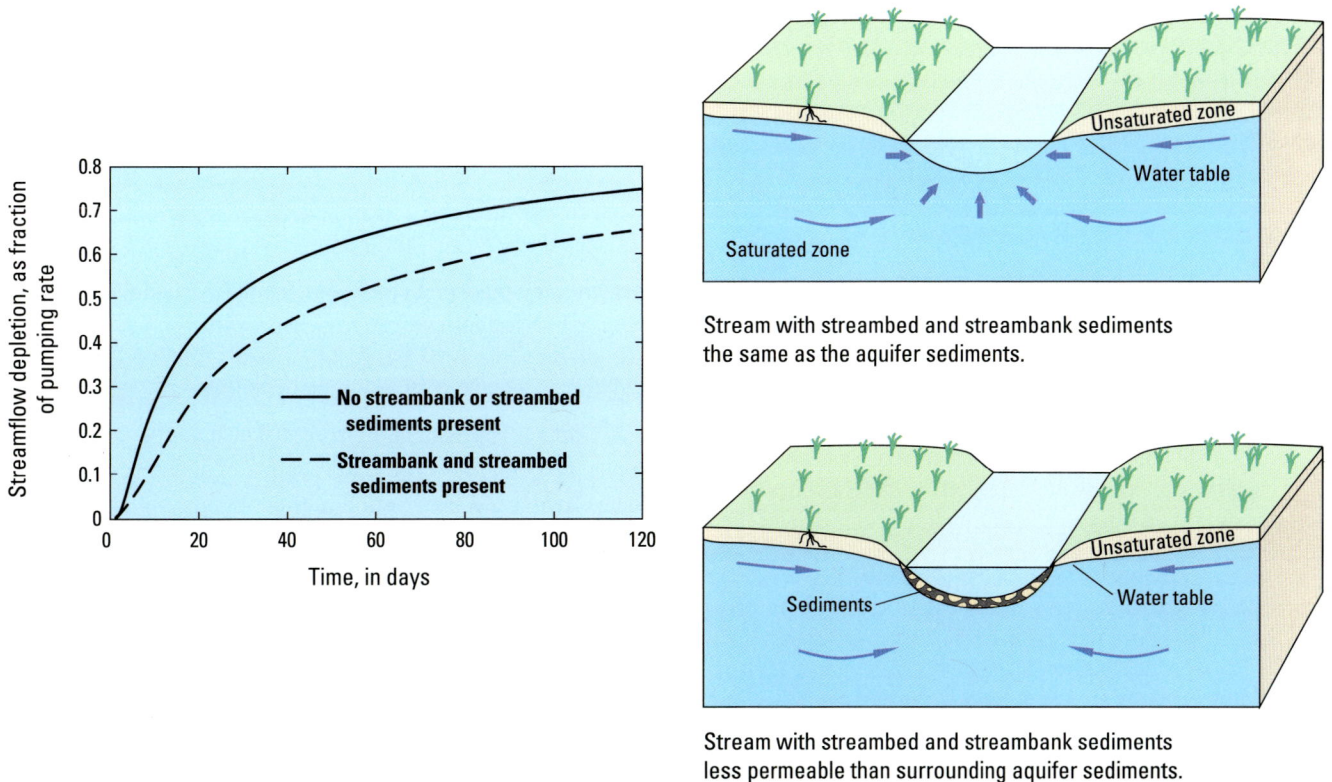

Stream with streambed and streambank sediments the same as the aquifer sediments.

Stream with streambed and streambank sediments less permeable than surrounding aquifer sediments.

Figure 11. Streamflow depletion resulting from a well pumping 500 feet from a stream at a rate of 250 gallons per minute. The presence of streambed and streambank materials with lower permeability than the surrounding aquifer reduces the amount of streamflow depletion during the 120 days of pumping. [Rates of streamflow depletion were calculated by using a computer program described in Reeves (2008); hydraulic diffusivity of aquifer is 10,000 feet squared per day and streambed leakance, which represents resistance between the stream and aquifer, is 200 feet.]

Box B: Ways to Express Streamflow Depletion

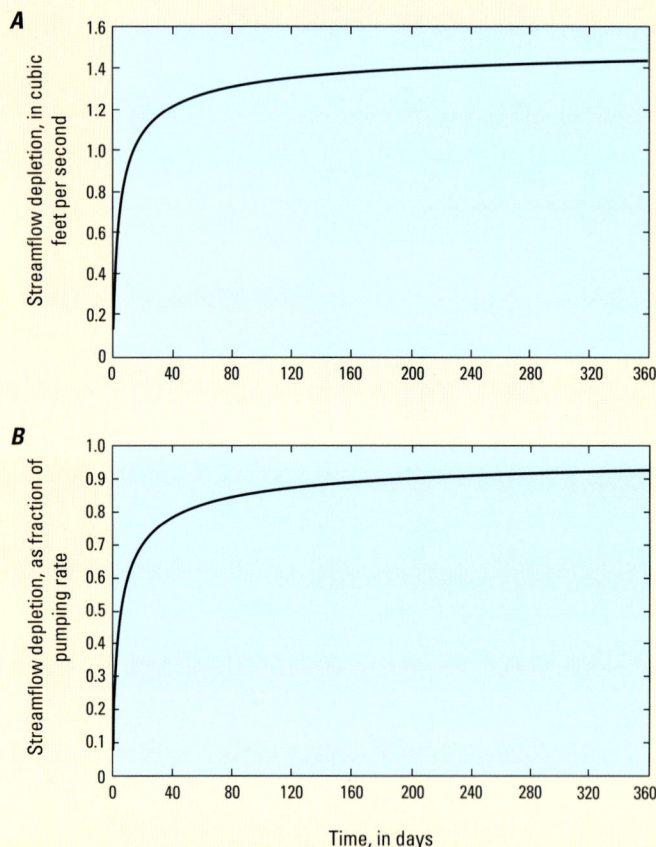

Figure B–1.   Streamflow depletion resulting from pumping at a well located 250 feet from a stream. The well is pumped at a rate of 1 million gallons per day (about 1.55 cubic feet per second). In graph *A*, streamflow depletion is expressed as a rate, in cubic feet per second; in graph *B*, depletion is expressed as a fraction of the pumping rate at the well, which is a dimensionless quantity. [Rates of streamflow depletion were calculated by using a computer program described in Reeves (2008); hydraulic diffusivity of the aquifer is 10,000 feet squared per day.]

Different approaches are used to quantitatively express the effects of groundwater pumping on streamflow. Some of these approaches are described and illustrated here to provide background for the discussions in the remainder of the report.

Change in streamflow rate

The most common way to describe streamflow depletion has been to report the change in the instantaneous flow rate of the stream, which is expressed in units of volume of streamflow per unit of time, such as cubic feet per second (ft^3/s), million gallons per day (Mgal/d), or acre-foot per year (acre-ft/yr). A related approach is to report the rate of streamflow depletion as a fraction of the pumping rate of the well, which is a dimensionless quantity.

These two approaches are illustrated in figure B–1, where rates of streamflow depletion are shown for a pumping rate of 1.0 Mgal/d at a well located 250 feet from a stream. The streamflow depletion that results from pumping the well is shown in units of cubic feet per second, which is the unit most often used in reporting streamflow. In these units, the pumping rate of the well is 1.55 ft^3/s, and the rate of streamflow depletion caused by pumping at the well is shown in the top graph of figure B–1 to approach this rate asymptotically. The bottom graph shows streamflow-depletion rates as a fraction of the pumping rate at the well for the same pumping conditions. In this case, the reporting unit is dimensionless, and the curve on the graph asymptotically approaches a value of 1.0.

Cumulative volume of streamflow depletion

Another approach used to describe streamflow depletion is the cumulative (or total) volume of streamflow that occurs over a specified period of time. In this approach, the units used are volumes of streamflow that are depleted, such as cubic feet, million gallons, or acre-feet. Because rates of streamflow depletion change over time, in order to calculate the total volume depleted over a period of time it is necessary to sum the volumes of depletion that occur over shorter time intervals within the full period of interest. For example, if one wants to determine the total volume of depletion over a 1-year period, an approach would be to sum the individual volumes of depletion that occur each day. These volumes could be calculated by multiplying the daily rates of streamflow depletion by the 1-day time interval. Volumes of depletion also can be expressed as a dimensionless fraction of the total volume of water pumped over the period of interest.

These concepts are illustrated in figure B–2 for the same pumping conditions described for figure B–1. Because the pumping rate at the well is constant at 1.0 Mgal/d, the cumulative amount of groundwater pumped increases linearly with time and is equal to 360 Mgal at the end of the 360-day pumping period (top curve in fig. B–2A). However, as shown in the graph, at any particular time, the total volume of streamflow depletion is less than the total volume of water pumped because of the delayed effect of the response of the stream to pumping at the well. The volume of streamflow depletion as a dimensionless fraction of total groundwater pumped is shown in the bottom graph of figure B–2. It should be noted that the dimensionless curve shown in the bottom graph of figure B–2 is not equal to the dimensionless curve in figure B–1 because the underlying responses (that is, rates of streamflow depletion as opposed to volumes of streamflow depletion) are different.

A closer look at where depletion occurs

Some situations may require detailed analyses of individual stream reaches subject to depletion. This is particularly true if depletion-related changes in water chemistry or temperature are of concern or if a goal is to maintain a minimum base flow in a critical stream reach. For these analyses, depletion can be reported as the instantaneous rate of depletion per unit length of stream, such as in units of cubic feet per second per mile. Detailed reach-by-reach estimates of depletion can be calculated with numerical groundwater-flow models to gain insight into where pumping-induced infiltration from the stream to the aquifer might occur. An example of such an analysis is shown in figure 15A.

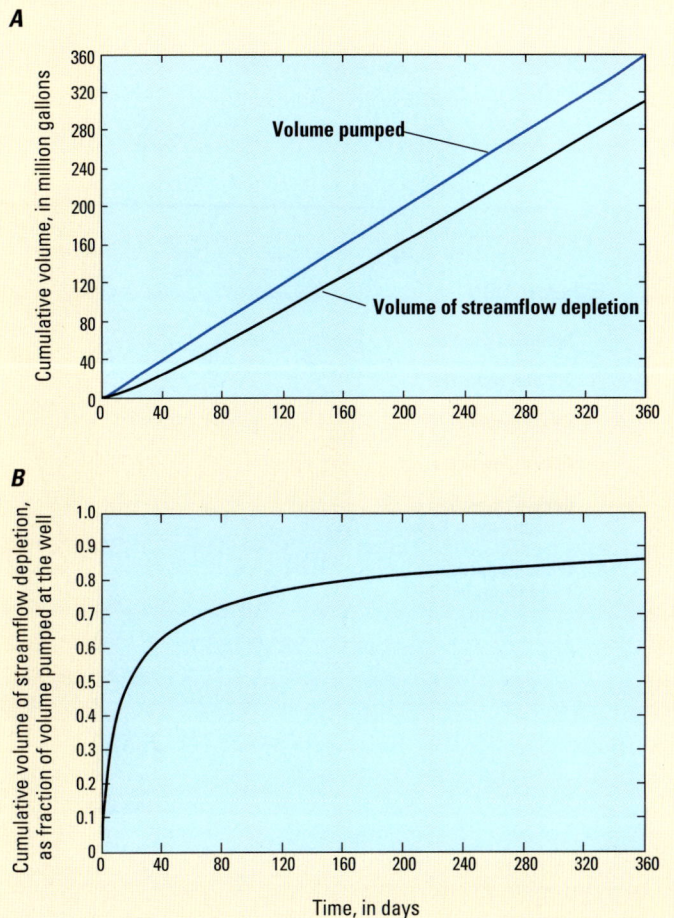

Figure B–2. Cumulative volume of streamflow depletion resulting from pumping at a well located 250 feet from a stream. The well is pumped at a rate of 1 million gallons per day. In graph A, streamflow depletion is expressed as the total (cumulative) volume of depletion that has occurred since the initiation of pumping, in million gallons; in graph B, the cumulative volume of depletion is expressed as a fraction of the cumulative volume of groundwater pumped at the well, which is a dimensionless quantity. [Volumes of streamflow depletion were calculated from the rates of streamflow depletion shown in figure B–1.]

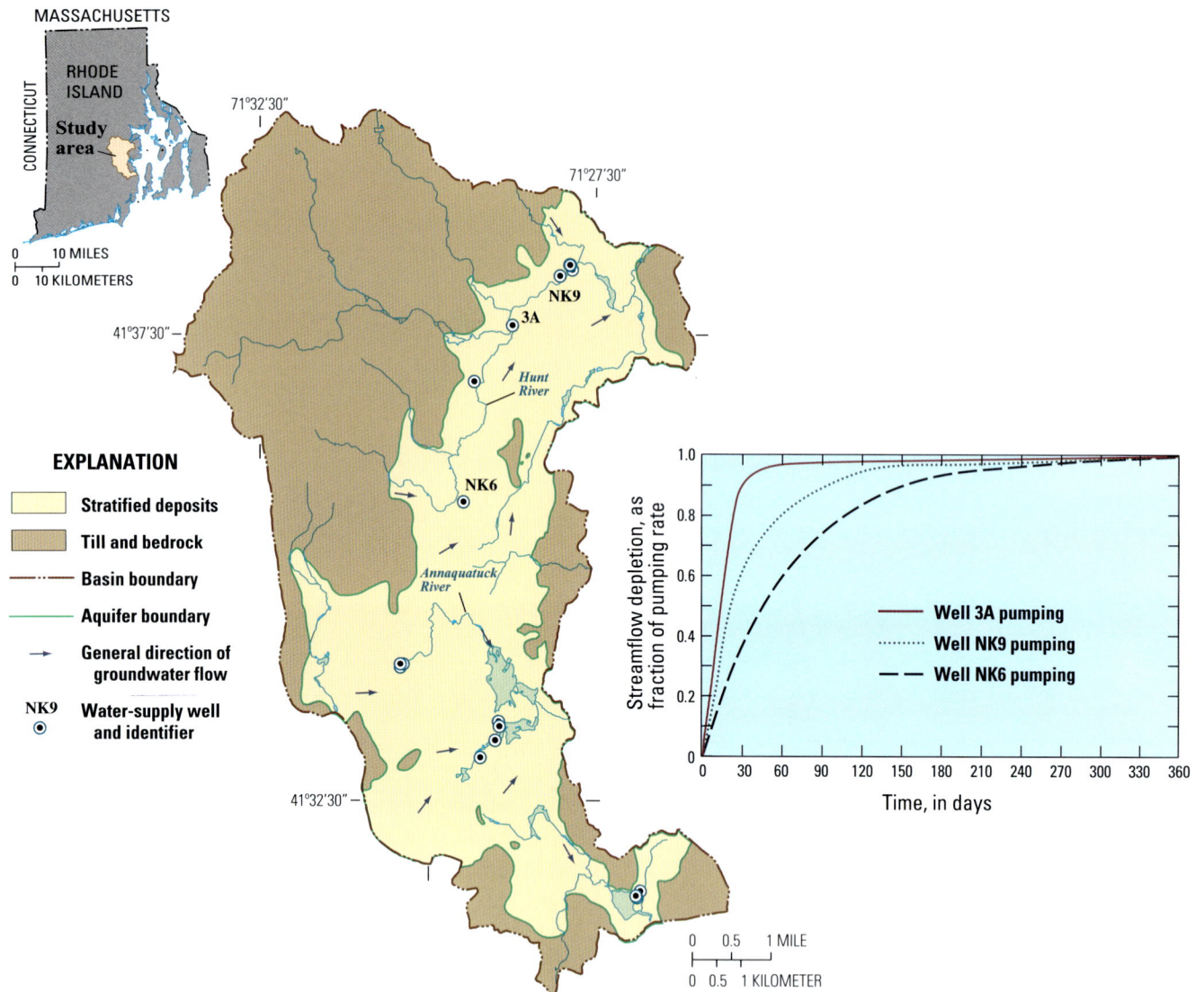

Figure 12. Streamflow depletion for three wells pumping from the unconfined-aquifer system of the Hunt River Basin, Rhode Island (map and streamflow-depletion data modified from Barlow and Dickerman, 2001).

of streamflow depletion computed for one pumping rate also can be applied to other pumping rates that do not cause substantial changes to the aquifer system. As noted in a later section of the report, the timing of depletion is also independent of rates and directions of groundwater flow in most aquifers. This means that depletion as a net effect on streamflow is the same whether a stream is gaining or losing, that features such as groundwater flow lines and divides have no influence on depletion, and that transient events such as changes in river stage or rates of aquifer recharge do not affect the timing of depletion by a pumping well (Leake, 2011).

Case studies from aquifers in the Eastern and Western United States illustrate the large differences in the timing of streamflow depletion that result from the variability in the scale of the two aquifer systems, proximity of the pumping wells to streams, and differences in the geology and hydraulic properties of the two aquifers. The case study from the Eastern

United States is the 40 square mile (mi²) stream-aquifer system of the Hunt River Basin of Rhode Island (fig. 12). The aquifer is typical of many of the glacially derived aquifers of the Northeastern United States that consist of stratified, unconsolidated sand and gravel sediments that are hydraulically connected to shallow streams, lakes, and ponds. The sediments were deposited by glacial meltwater within generally narrow river valleys bounded laterally and at depth by glacial till and bedrock. The sand and gravel deposits can have very high values of transmissivity, even though they are often no more than 100 to 150 ft thick at the deepest part of the valley. The aquifers typically are unconfined and have substantial storage capacities. Water-supply wells frequently are placed close to the streams where the valley depth and aquifer transmissivity are greatest. Thus, the distance from the wells to the groundwater-discharge boundaries at the streams is often less than a few hundred feet.

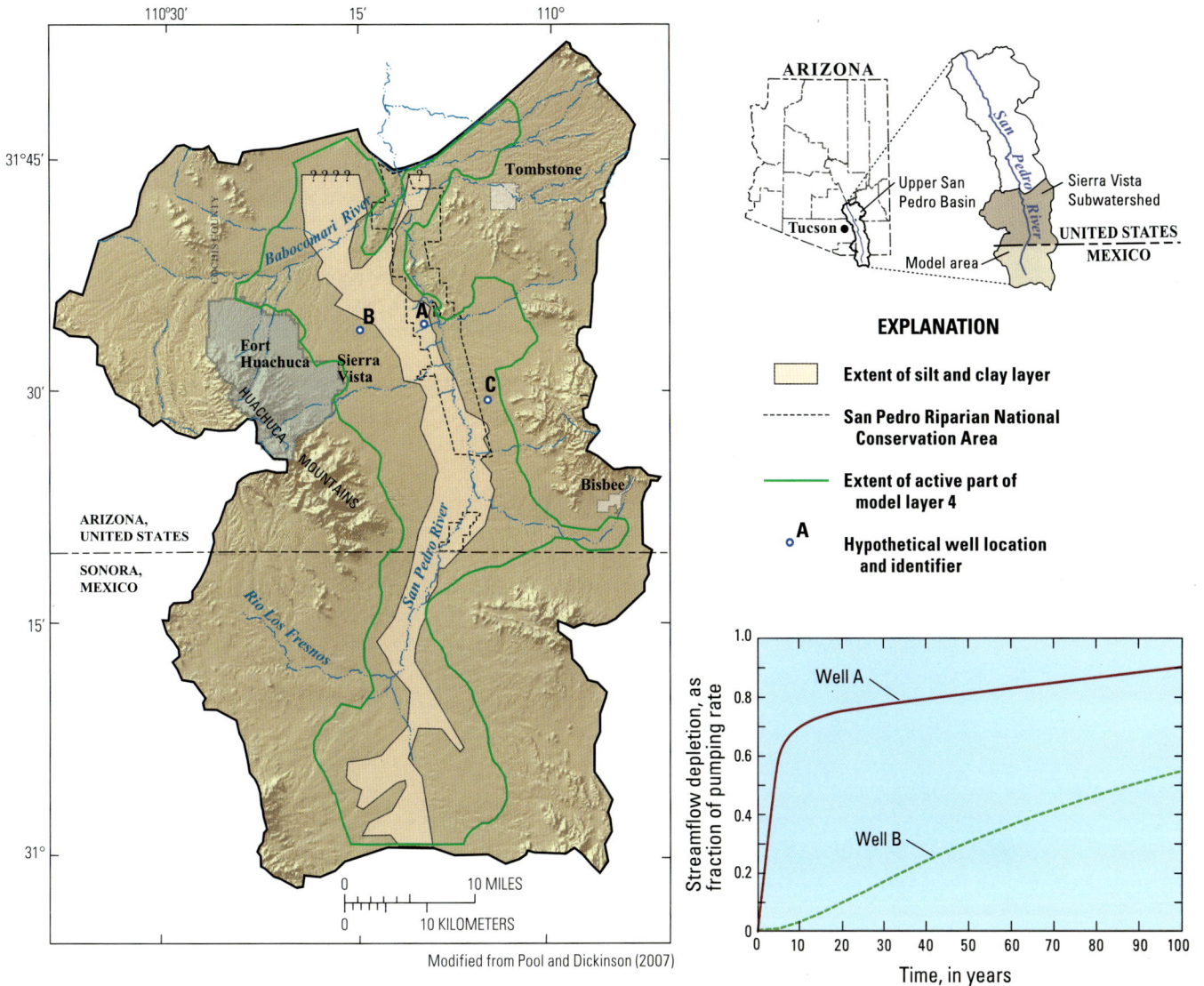

Figure 13. Streamflow depletion for hypothetical well locations A and B pumping from the Upper San Pedro Basin aquifer system, southern Arizona (modified from Leake, Pool, and Leenhouts, 2008.) [Well C is discussed later in the report.]

The distribution of water-supply wells in the Hunt River Basin is typical of wells in these river-valley aquifers (fig. 12). The majority of the wells are clustered along the Hunt and Annaquatucket Rivers, where the transmissivities of the aquifer are largest (Barlow and Dickerman, 2001). Nearly all of the wells are within about 500 ft of a stream from which groundwater that would otherwise have discharged to the stream is captured or streamflow is drawn into the aquifer by the process of induced infiltration. Because of the close proximity of the wells to the streams and the relatively high transmissivity of the aquifer near the wells, the time response of streamflow depletion to pumping is relatively fast, as illustrated by the streamflow-depletion curves calculated by a numerical groundwater-flow model of the basin for three wells that pump near the Hunt River (fig. 12). Each of the three wells captures more than 90 percent of its withdrawal from streamflow depletion within 180 days of the start of pumping,

and the time to reach a depletion-dominated supply is less than 50 days for each well.

In contrast to the narrow and relatively shallow alluvial-valley aquifer settings of the Northeast, many aquifers of the Western United States extend over hundreds to thousands of square miles and are hundreds of feet thick. An example groundwater system of the West is that within the Upper San Pedro Basin that extends from northern Sonora, Mexico, into southern Arizona (fig. 13). The watershed covers an area of about 1,700 mi^2. Groundwater discharge sustains perennial reaches in the San Pedro River and tributaries, as well as narrow bands of groundwater-dependent vegetation adjacent to streams. The riparian area provides year-round habitat for aquatic and terrestrial wildlife species, and also is an important corridor for birds migrating between Mexico and the United States. The San Pedro Riparian National Conservation Area, managed by the Bureau of Land Management, was established in 1988 to protect and enhance this desert ecosystem.

Photograph by Bob Herrmann

The primary aquifer within the Upper San Pedro Basin comprises thick alluvial deposits that occupy a structural basin that lies between rocks in the surrounding mountains (Pool and Dickinson, 2007; Leake, Pool, and Leenhouts, 2008). The basin-fill deposits are subdivided into upper, highly permeable and lower, less permeable parts that are collectively as much as 1,700 ft thick. An extensive silt and clay layer that vertically spans parts of the upper and lower basin fill separates the aquifer into deep confined and shallow unconfined sections. The areal extent of the silt and clay layer is shown in figure 13. Groundwater within the Upper San Pedro Basin generally flows from recharge areas near the mountains to areas near the San Pedro River where it discharges to the stream and springs or is evaporated or transpired by riparian vegetation. A portion of the groundwater flow is intercepted upgradient from the streams by pumped wells.

Graphs of streamflow depletion for two wells pumping within the San Pedro Basin in figure 13 illustrate the large difference in response times for pumping from this system compared to that of the Hunt River Basin. Response times for streamflow depletion in the San Pedro River Basin are measured in years and decades, whereas those for the Hunt River are measured in days and months. For example, in contrast to wells pumping in the Hunt River Basin, the time required to reach depletion-dominated supply is about 5 years for well A and nearly 90 years for well B. The long response times for the San Pedro Basin result from the relatively large distances of hypothetical wells A and B from the San Pedro River (about 1.5 miles (mi) for well A and 6 mi for well B) and the specific characteristics of the groundwater system of the San Pedro River Basin, including its large areal extent, the thickness of the basin-fill sediments, and the presence of the silt and clay confining unit, all of which increase the time during which the wells draw from aquifer storage. These long response times have implications to monitoring and managing streamflow depletion in these aquifer settings—a topic that will be discussed later in the report.

The discussion in this section on the time response of streamflow depletion to pumping is based on concepts of the source of water to pumped wells that have been presented in literature over a period of more than 70 years, including those by Theis (1940), Lohman (1972), Bredehoeft and others (1982), Heath (1983), Alley and others (1999), Bredehoeft (2002), and Bredehoeft and Durbin (2009). These concepts are also relevant to the related topics of groundwater budgets and groundwater sustainability.

San Pedro River below Hereford, Arizona. The riparian zone along the river provides abundant food, water, and cover for hundreds of species of birds, including the Vermilion Flycatcher. (Background photograph by Michael Collier)

Distribution of Streamflow Depletion Along Stream Reaches

The cone of depression that forms around a well extends outward in all directions from the well, and, as a result, groundwater pumping affects streams and stream reaches that are both upgradient to and downgradient from the location of withdrawal. Some stream reaches will be affected more than others, depending on the distance of the well from the reach and the three-dimensional distribution and hydraulic properties of the geologic materials that compose the groundwater system and adjoining streambeds. Steep hydraulic gradients at the stream-aquifer interface created by the pumping may cause some stream reaches to become losing, while other reaches remain gaining. Streamflow depletion increases in the downstream direction of a basin, and if depletion is the only source of water to the pumped well, the rate of depletion over time will tend to approach the pumping rate of the well in the direction of the outflow point (or points) of the basin.

These concepts are illustrated by the results of numerical simulations for two different aquifers, the first a hypothetical groundwater system representative of river-valley aquifers of the Northeastern United States and the second a real-world aquifer in northern Arizona. The hypothetical system consists of a single stream that receives groundwater discharge along its entire length from a hydraulically connected unconfined aquifer (fig. 14; based on Barlow, 1997). The unconsolidated sand and gravel deposits that fill the valley range in saturated thickness from about 40 ft along the boundaries of the valley to a maximum of about 100 ft in the center of the valley. For simplicity, the hydraulic properties of the aquifer and streambed sediments are homogeneous. The aquifer is bounded at depth by impermeable bedrock. The primary source of water to the groundwater system is recharge at the water table, but groundwater also flows into the river valley from the surrounding uplands and along the northern boundary of the simulated area. Groundwater leaves the system primarily as discharge to the stream but also along the southern boundary of the simulated area.

The rate of groundwater discharge to the stream is constant along its entire length in the absence of pumping (fig. 15A), which results in a linear increase in streamflow from zero cubic foot per second (ft³/s) at the upstream end of the simulated basin to 6.1 ft³/s at the outflow point of the basin (fig. 15B). The uniform rate of groundwater discharge to the stream results from the symmetry of the system and assumed homogeneity of the hydraulic properties of the aquifer and streambed materials.

The effects of pumping at three well locations are evaluated for steady-state flow conditions; that is, for conditions in which groundwater levels are no longer declining, aquifer-storage depletion is no longer occurring, and streamflow depletion is the only source of water to the wells. The three wells are located midway between the northern and southern boundaries of the system at distances of 300 ft (well A), 700 ft (well B), and 1,400 ft (well C) from the stream (fig. 14).

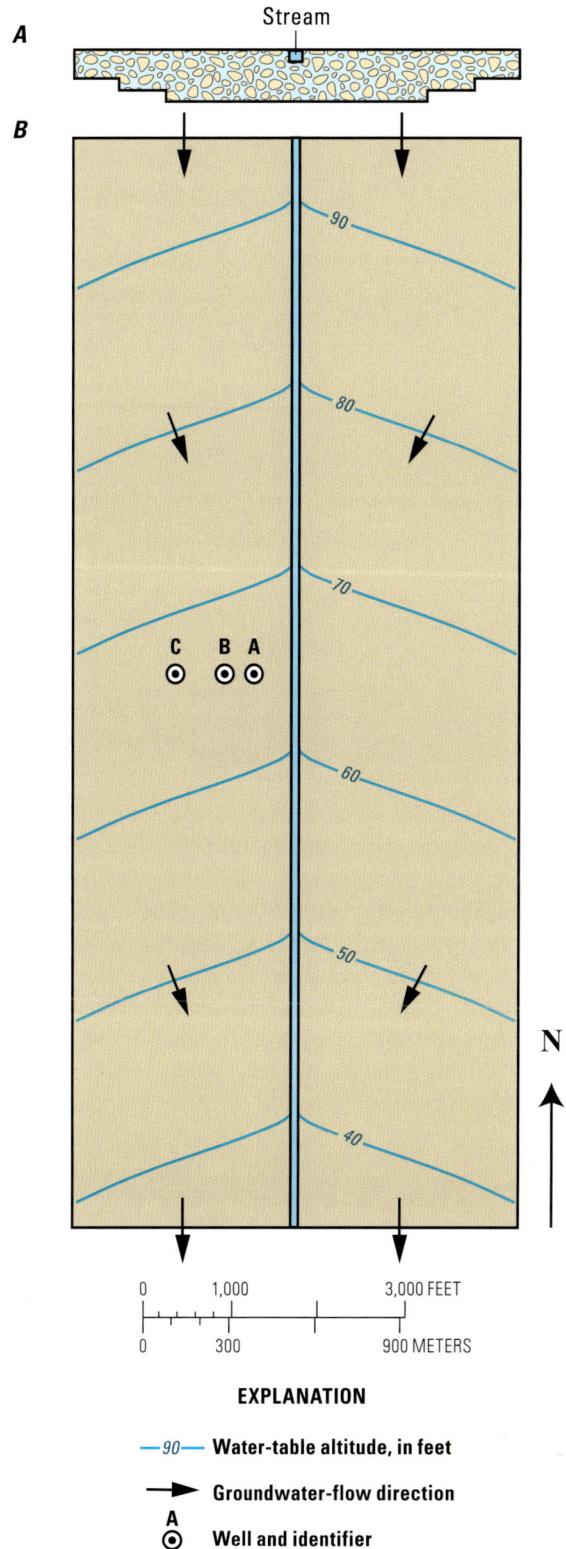

Figure 14. *A*, Cross section of a hypothetical river-valley aquifer with a shallow stream. *B*, Plan view of the water-table altitude and groundwater-flow directions in the aquifer with no pumping at the three wells. The stream receives groundwater discharge along its entire length (modified from Barlow, 1997).

A

B

EXPLANATION

— **Well A pumping**
— **Well B pumping**
— **Well C pumping**

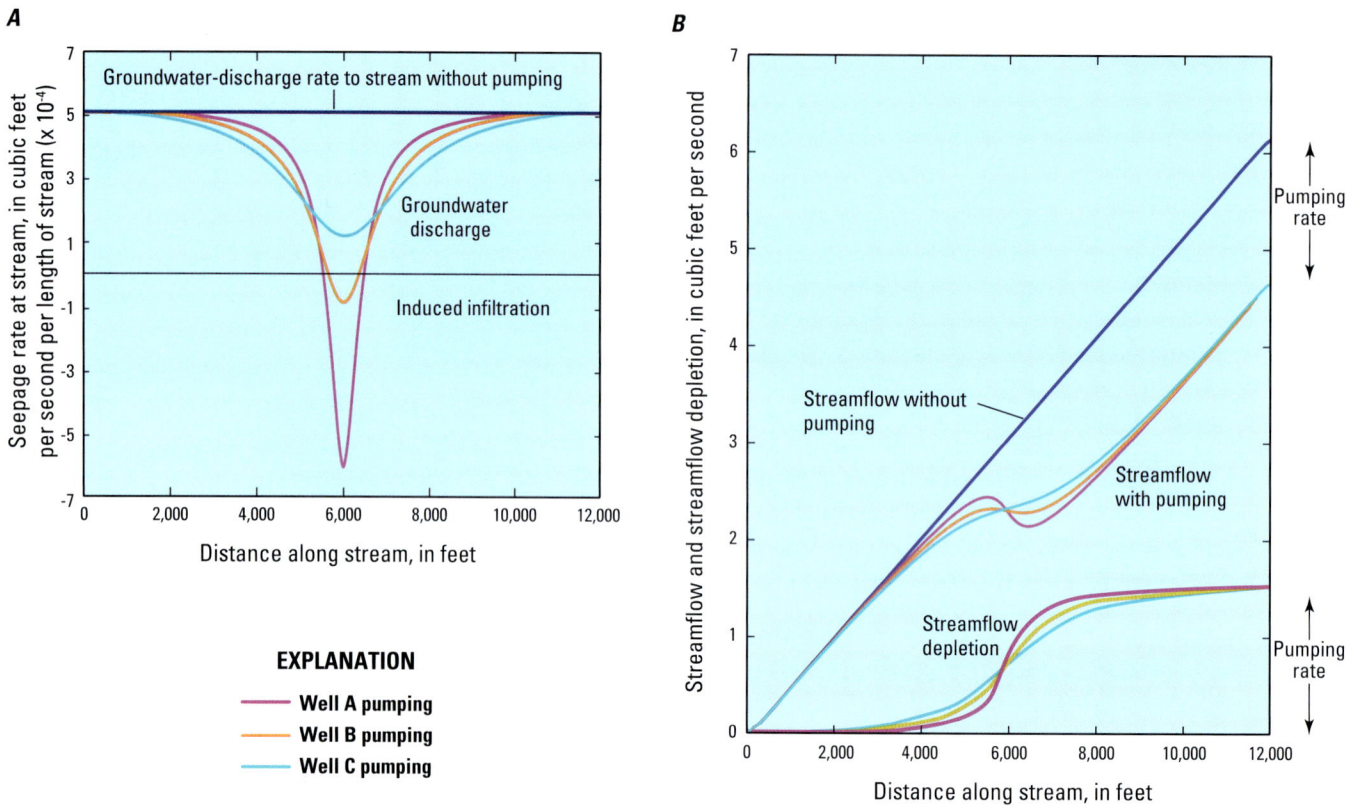

Figure 15.   *A*, Rates of groundwater discharge (positive values) and induced infiltration (negative values) along the 12,000-foot reach of the stream shown in figure 14*B*: In the absence of pumping, the stream receives groundwater discharge at a uniform rate along its entire length. Pumping from wells A and B, which are located 300 feet and 700 feet from the stream, respectively, cause induced infiltration along part of the stream, whereas pumping at well C, which is located 1,400 feet from the stream, does not. Reaches that are gaining coincide with locations of groundwater discharge, whereas reaches that are losing coincide with locations of induced infiltration. *B*, Streamflow and streamflow depletion along the stream: In the absence of pumping, there is a linear increase in streamflow along the entire stream length. With pumping, streamflow depletion increases in the downstream direction and approaches the pumping rate of each well (1.55 cubic feet per second), regardless of the distance of each well from the stream. The results shown in these graphs are for steady-state flow conditions. Well locations are shown in figure 14. (Results from models documented in Barlow, 1997.)

Each well is pumped independently of the others at a rate of 1 million gallons per day (Mgal/d; 1.55 ft³/s) in three separate simulations.

The graph in figure 15*A* shows the distribution and rates of streambed seepage along the stream for pumping at the three wells. Seepage rates greater than zero indicate groundwater discharge to the stream and gaining streamflow conditions, whereas seepage rates less than zero indicate induced infiltration of streamflow into the aquifer and losing streamflow conditions. The graph indicates that changes in streambed seepage rates are not confined to the reach of the stream that is immediately opposite to the wells (that is, at 6,000 ft along the stream), but instead extend both upgradient to and downgradient from the wells. The graph also indicates that induced-infiltration rates are largest for the well closest to the stream (well A) and decrease as the distance of the pumped well from the stream increases.

The graphs in figure 15*B* demonstrate that streamflow depletion also occurs in both the upgradient and downgradient directions from the pumped wells and that the amount of depletion asymptotically approaches the pumping rate of each well (1.55 ft³/s) in the downstream direction, regardless of the distance of the pumped well from the stream. This results from the fact that for the conditions simulated (including the steady-state flow conditions) there are no other sources of water to the wells. As a consequence, all of the water pumped by each well must come from streamflow depletion, either as captured groundwater discharge or induced infiltration.

The change in the groundwater-discharge rates to the stream and resulting changes to streamflow that occur in response to pumping at the wells can be explained by the distribution of groundwater-level declines (drawdowns) that form around each pumped well (fig. 16). Because well A is closest to the stream, the drawdowns created at the well are larger

A   B

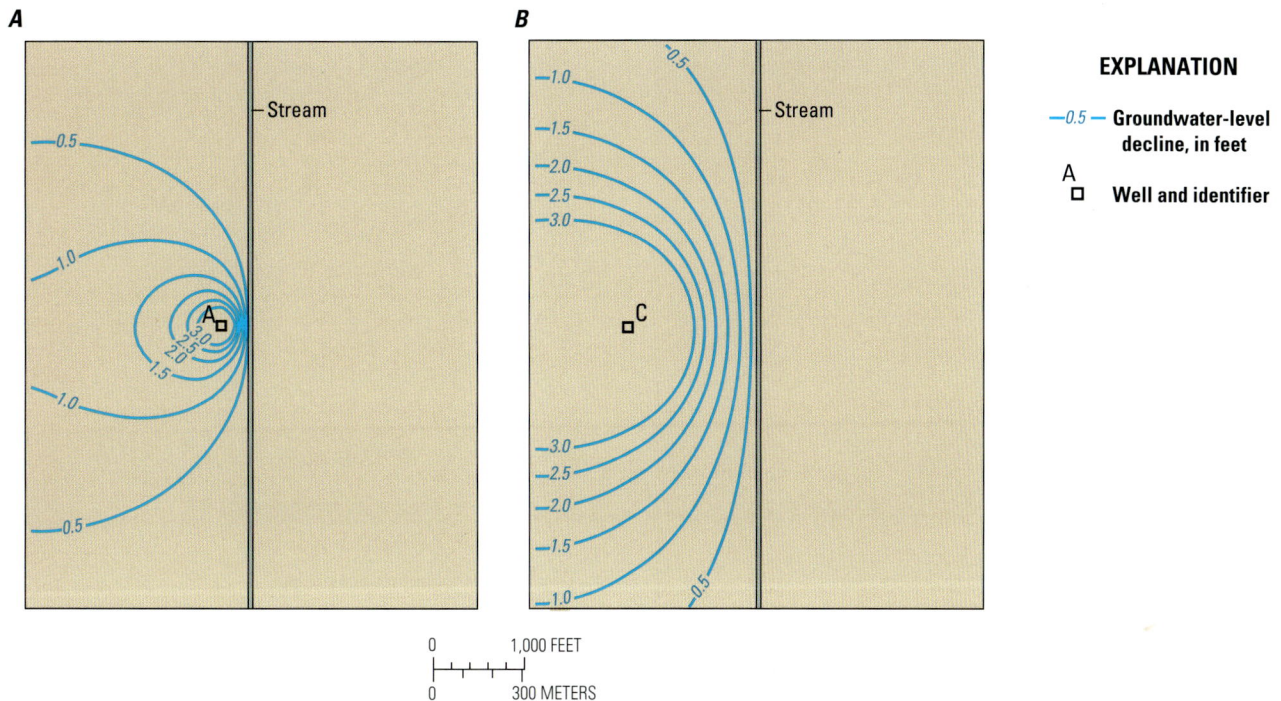

Figure 16. Groundwater-level declines (drawdowns) in the aquifer caused by, *A,* pumping at well A, and, *B,* pumping at well C. Well locations are shown in figure 14. (Results from models documented in Barlow, 1997.)

at the stream-aquifer boundary than those that are created by pumping at well C. As a result, the hydraulic gradients at the stream boundary and resulting rates of induced infiltration are largest for pumping at well A and smallest for pumping at well C. The symmetry of the seepage and streamflow-depletion rates shown in figure 15 results from the symmetry of the drawdown curves around each well, which results from the straight boundaries along the modeled aquifer and of the stream and the uniformity of the simulated aquifer sediments.

The real-world example considered here is from a study of the C aquifer in northeastern Arizona. Groundwater discharge from the C aquifer supports threatened and endangered fish species in some reaches of the Little Colorado River and its tributaries (Leake and others, 2005). The aquifer is named for the Coconino Sandstone, which is the principal water-bearing unit of the aquifer. The primary discharge area for the C aquifer in the Little Colorado River Basin is a series of springs in the lower reaches of the Little Colorado River, including Blue Spring. Discharge also occurs along various reaches of the tributaries to the Little Colorado River, such as the lower reaches of Clear and Chevelon Creeks (fig. 17); base-flow measurements made around the time of the C-aquifer study indicate groundwater discharge rates of about 5.6 and 2.7 ft³/s, respectively, to the two reaches.

The C aquifer is the most productive aquifer in the Little Colorado River Basin and an important source of water for many users (Leake and others, 2005). Increased withdrawals

from the aquifer have been proposed from a cluster of wells just south of Leupp, Arizona, to meet increased future demands (fig. 17). A numerical model of change in groundwater flow in the aquifer was developed to evaluate the potential effects of future withdrawals on groundwater discharge to perennial reaches in the Chevelon and Clear Creek drainages and to the Little Colorado River in the area below Blue Spring (fig. 17). Two withdrawal scenarios were simulated for a 101-year period that included 51 years of withdrawals followed by 50 years of no withdrawals. Scenario A simulates a nearly constant withdrawal rate of about 9.0 ft³/s [about 6,500 acre-feet per year (acre-ft/yr)], and scenario B simulates a more complicated pumping pattern with a maximum withdrawal rate of about 15.9 ft³/s (11,500 acre-ft/yr; fig. 18).

The time distribution of streamflow depletion is shown for all stream reaches in figure 19A and for only the lower Clear and Chevelon Creek reaches in figure 19B. All of the depletion curves shown in the figure indicate a gradual, steady increase in the streamflow-depletion rates for both scenarios, even though the withdrawal schedule for scenario B is quite variable. This damping of the pumping variability results from the large distances of the wells from the stream reaches of interest and from the diffusivity of the aquifer. Total streamflow depletion for all reaches at the end of the 51-year pumping period is about 0.31 ft³/s for scenario A and 0.37 ft³/s for scenario B (fig. 19A). Of the total depletion shown in figure 19A for the two scenarios, nearly all of the

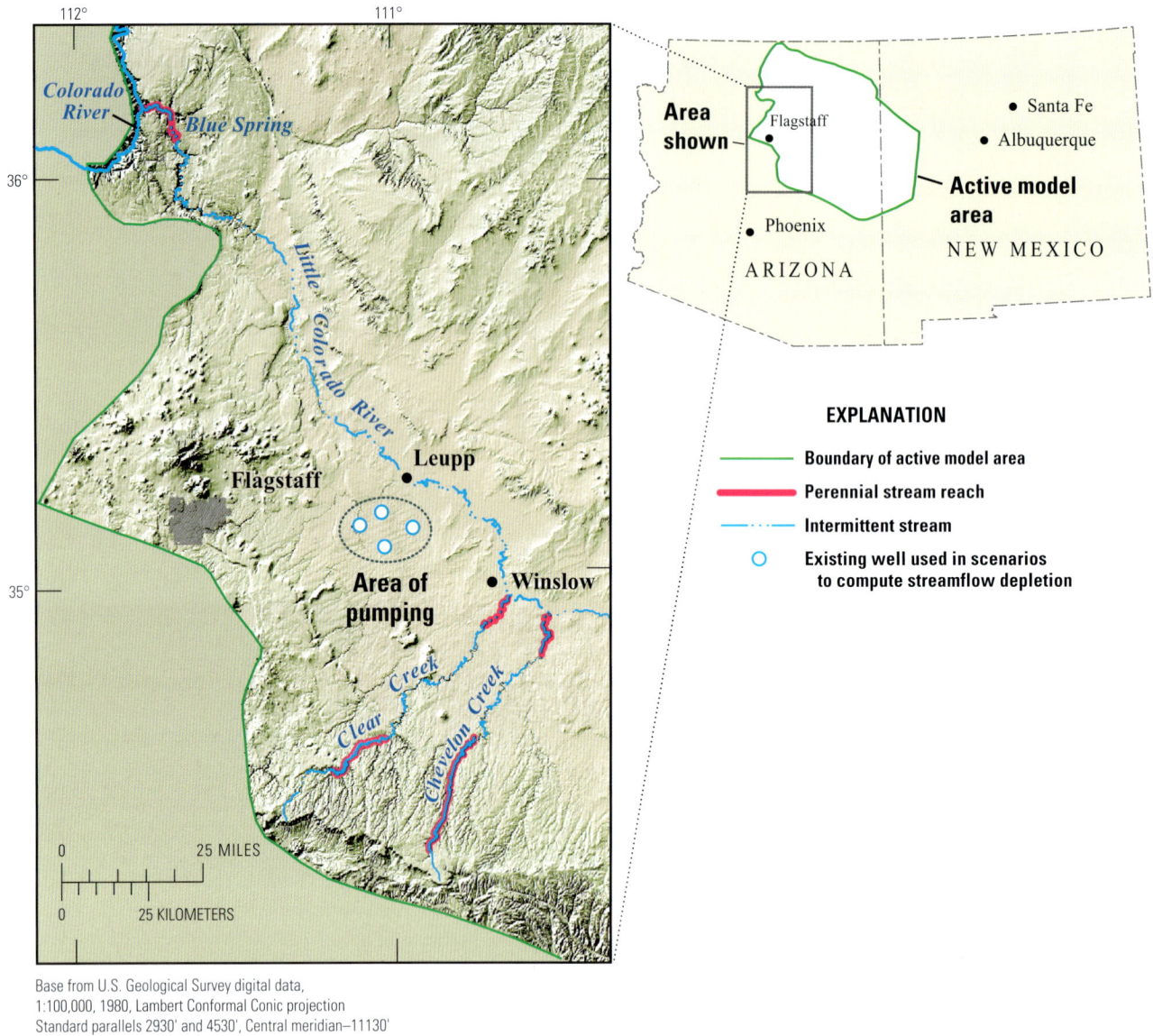

Base from U.S. Geological Survey digital data,
1:100,000, 1980, Lambert Conformal Conic projection
Standard parallels 2930' and 4530', Central meridian–11130'

Figure 17.   Locations of perennial stream reaches and some of the wells simulated in proposed withdrawal scenarios for the C aquifer, northeastern Arizona (modified from Leake and others, 2005).

depletion occurs in the two reaches that are closest to the pumping center, lower Clear Creek, which is about 25 mi from the wells, and lower Chevelon Creek, which is about 34 mi from the wells (fig. 19*B*).

The results shown in figure 19 indicate that the rates of depletion in all stream reaches after 51 years of pumping are relatively small compared to the maximum pumping rates for each scenario. For example, for scenario A, streamflow-depletion rates are 0.26 ft³/s for lower Clear Creek and 0.05 ft³/s for lower Chevelon Creek at the end of the pumping period (fig. 19*B*), yet the maximum pumping rate at the well cluster is about 9.0 ft³/s for this scenario. These results indicate that nearly all of the water pumped by the wells during the pumping period is derived from reductions in aquifer storage and that the system is far from reaching a new steady-state condition.

The responses shown in the figure also indicate that streamflow depletion continues long after pumping stops. This results from the fact that the withdrawal locations are far from the perennial stream reaches—recovery from shutting off withdrawals takes time to reach distant parts of the outward-propagating cone of depression (Leake and others, 2005). Maximum streamflow-depletion rates for all reaches taken together (fig. 19*A*) occur at about year 95 for both withdrawal scenarios, about 44 years after pumping stops. For the lower Chevelon Creek reach, however, maximum depletion occurs even later in time, near the end of the 101-year simulation period (fig. 19*B*). Residual pumping effects on streamflow depletion are discussed in more detail in the section "Depletion after Pumping Stops" in the chapter on common misconceptions about streamflow depletion.

Figure 18. Withdrawal scenarios simulated for the C aquifer, northeastern Arizona (modified from Leake and others, 2005).

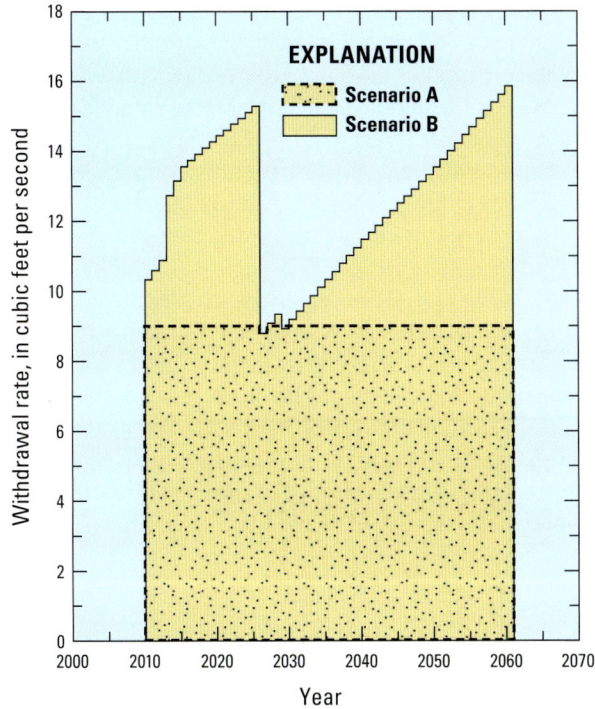

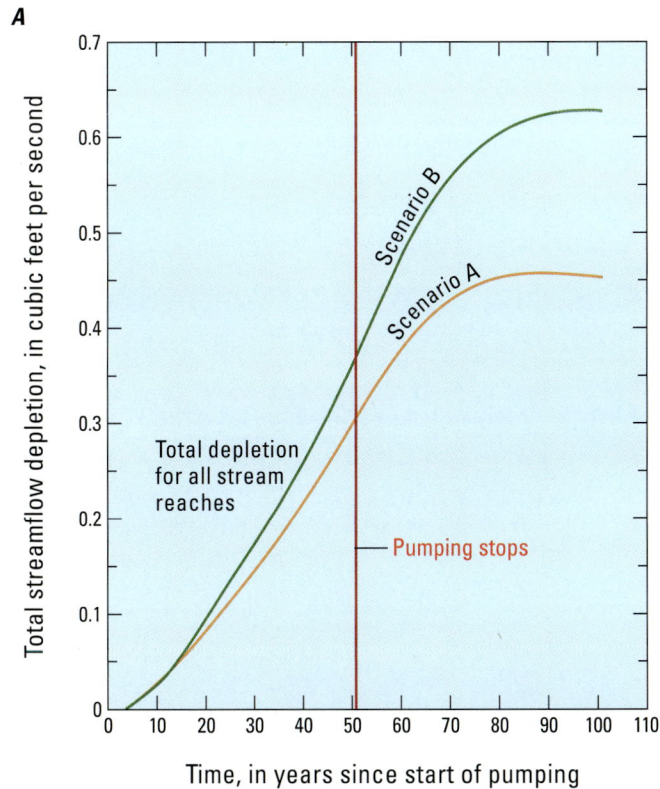

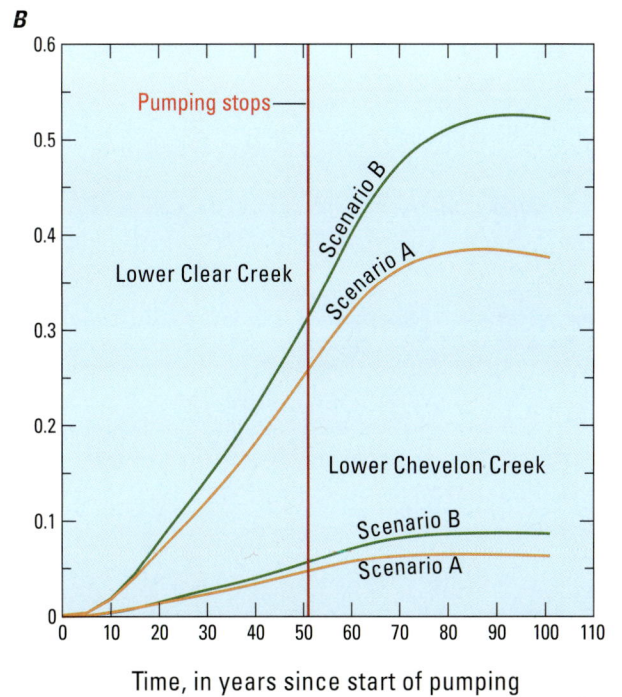

Lower Clear Creek near Winslow, Arizona. (Photograph by Robert J. Hart, U.S. Geological Survey)

Figure 19. Streamflow depletion as a function of time for two scenarios of groundwater pumping from the C aquifer, northeastern Arizona. *A,* All stream reaches. *B,* The lower Clear and Chevelon Creeks. Withdrawals in both scenarios stop at the end of year 51, indicated by the vertical red lines (modified from Leake and others, 2005).

Variable and Cyclic Pumping Effects

Previous sections of this report have focused primarily on streamflow depletion for conditions in which pumping occurs at a single rate for an extended period of time. More commonly, however, pumping schedules vary with time, either in response to changing water-supply demands or for maintenance and overall operation of the water-supply system. Pumping schedules can vary on hourly and daily bases in response to short-term fluctuations in demands and over longer-term cycles in response to factors such as seasonal and annual climate variability and irrigation demands. Two examples of the effects of variable and cyclic pumping on streamflow depletion are described for two different water-supply settings—a well that pumps to meet daily fluctuations in public-supply demands and one that pumps on a seasonal pumping and nonpumping cycle to meet irrigation demands.

The examples demonstrate that the overall effect of the diffusive properties of an aquifer are to dampen the variability and amplitude (range) of the pumping rates, such that the resulting rates of streamflow depletion are less variable and smaller in amplitude.

Groundwater withdrawals for primarily domestic and commercial uses in the Ipswich River Basin in eastern Massachusetts have caused substantial depletions of streamflow during summer low-flow periods (Zarriello and Ries, 2000). In the past, these depletions stressed aquatic communities and caused fish and mussel kills during dry years (Armstrong and others, 2001; Glennon, 2002). Pumping rates at one of the water-supply wells in the basin during a 9-year period illustrate the variability in withdrawals that occur in response to fluctuations in water-supply demands, which are generally highest during the spring and summer but then decrease during the fall and winter (fig. 20).

Photograph by David Armstrong, U.S. Geological Survey

Pool and dry river bed along the Ipswich River, Reading, Massachusetts, September 2005.

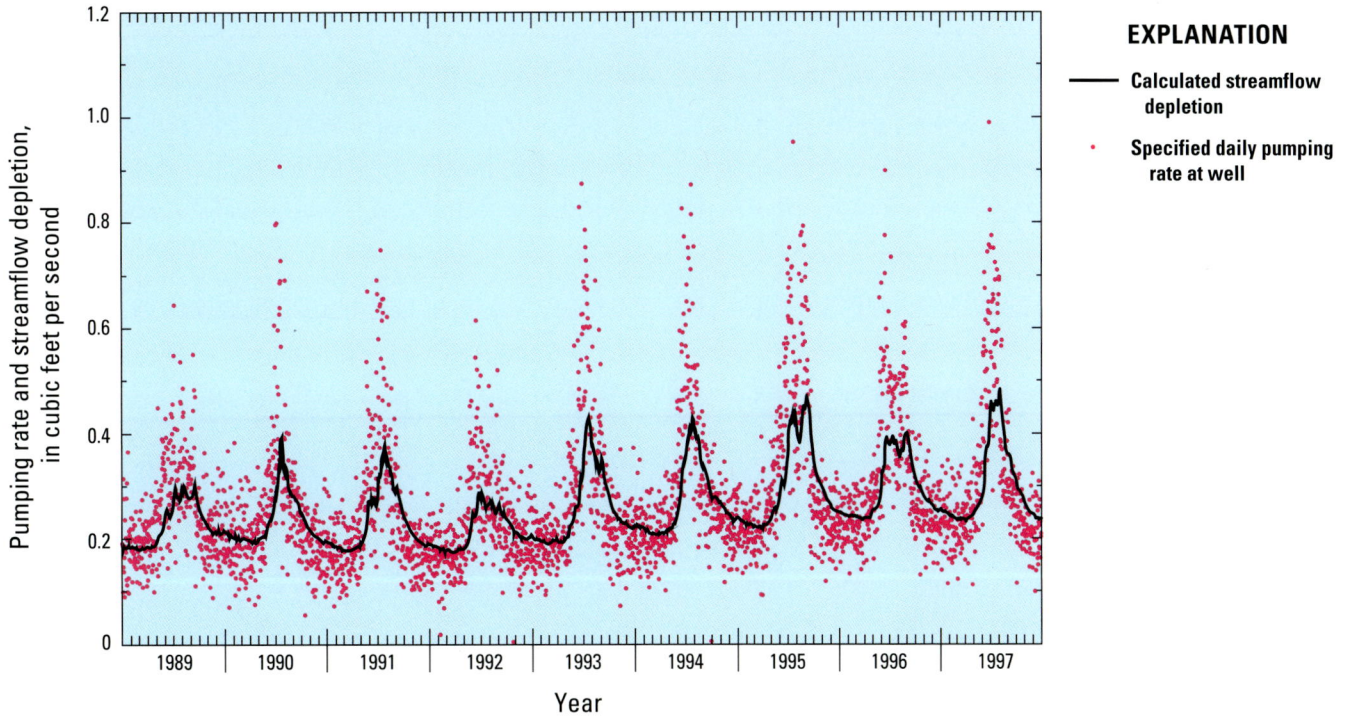

Figure 20. Daily pumping rates and model-calculated streamflow depletion for a well pumping about 500 feet from the Ipswich River, Massachusetts (modified from Barlow, 2000).

Rates of streamflow depletion in the Ipswich River that result from the daily pumping schedule also are shown in the figure and were calculated with an analytical model of streamflow depletion (Jenkins, 1968a; Barlow, 2000). As can be seen in figure 20, the range and variability of calculated streamflow-depletion rates are much less than those for the pumping rates at the well. The factors that cause reductions in the amplitude and variability of the pumping rates that are represented in the analytical model are the distance of the well from the river (about 500 ft) and the hydraulic diffusivity of the aquifer. Calculated streamflow depletion also exhibits an upward trend during the 9-year period, which is consistent with the upward trend in groundwater withdrawals.

Groundwater is a source of irrigation water to some of the most productive agricultural areas of the country, including California's Central Valley and the High Plains area of the Midwest. The hydrogeologic setting of many agricultural areas is often a river-valley system consisting of one or more streams that are in hydraulic connection with a shallow, unconfined aquifer. Pumping rates from these aquifers are largest during the irrigation season but then decrease sharply during the remainder of the year. The effect of cyclic pumping on streamflow depletion has been the subject of much research, and the remaining discussion draws on contributions by Jenkins (1968a), Wallace and others (1990), Darama (2001), Chen and Yin (2001), Kendy and Bredehoeft (2006),

Bredehoeft and Kendy (2008), and Bredehoeft (2011a). An example of the effects of cyclic pumping at a hypothetical agricultural supply well placed at various distances from a stream is described here and is similar to examples of cyclic pumping provided by Bredehoeft and Kendy (2008) and Bredehoeft (2011a).

The annual pumping cycle for the hypothetical agricultural supply well is illustrated in the top graph of figure 21. The well withdraws water from an extensive, unconfined aquifer that is bounded on one side by a single river. Pumping to meet irrigation demands occurs from June through August of each year; the irrigation season is then followed by a 9-month period of no withdrawals. The effects of this annual pumping cycle on streamflow depletion are illustrated for a 15-year period of pumping at three wells located at different distances from the river—300 ft, 1,000 ft, and 3,000 ft. The total volume of water pumped at the well during each irrigation season is 365 Mgal. When averaged over the full year, the withdrawal rate at the well is 1.0 Mgal/d (about 1.5 ft³/s), but because withdrawals only occur during the 3-month (92-day) period, the actual pumping rate is nearly 4 Mgal/d (about 6.1 ft³/s) during the irrigation season (fig. 21A). For comparison, streamflow depletions resulting from the cyclic pumping pattern are contrasted with those resulting from a constant pumping rate at the well for the entire 15-year period of 1.0 Mgal/d.

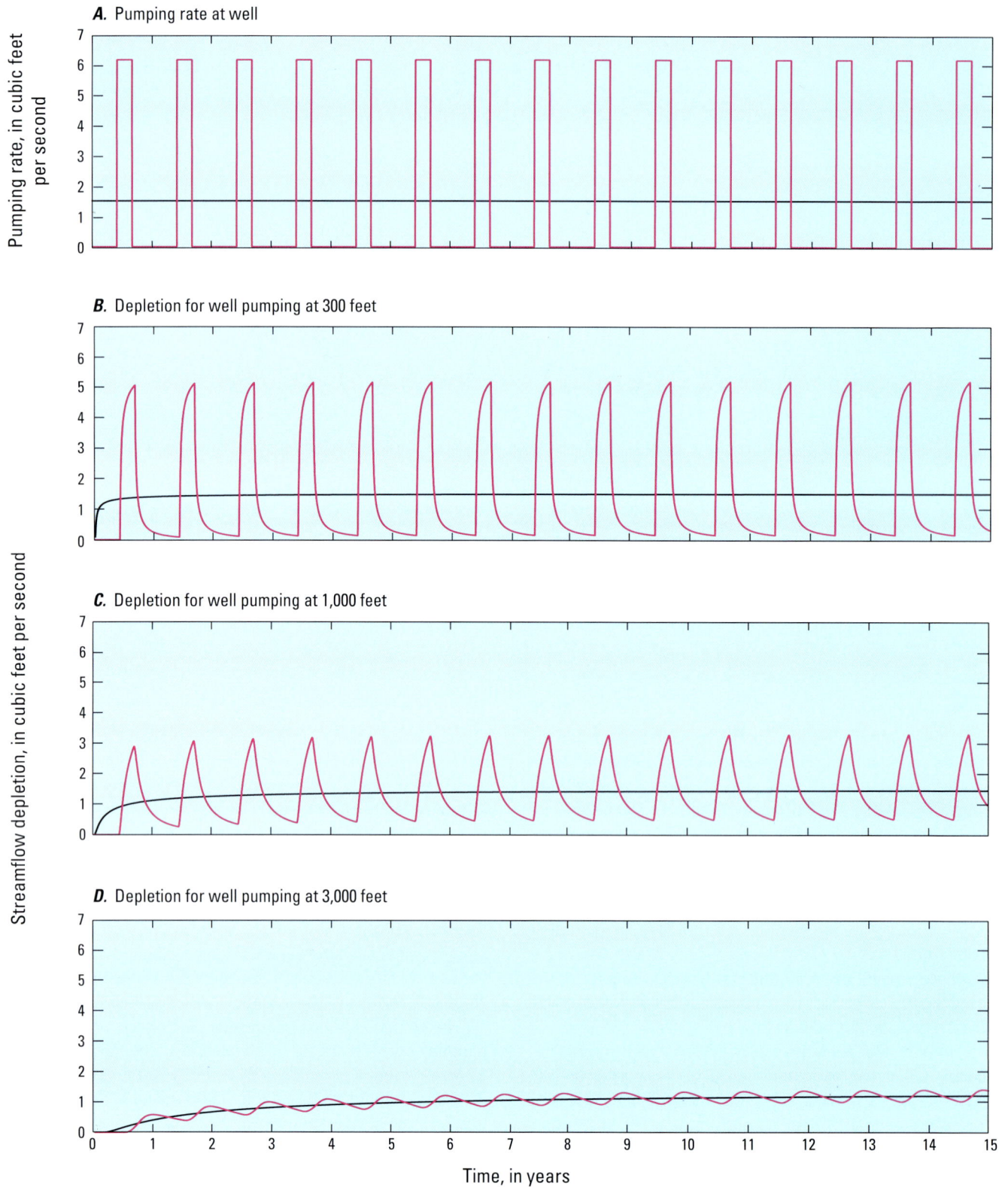

Figure 21. Patterns of streamflow depletion for both seasonal and constant pumping rates. *A,* The constant pumping rate, shown by the black line, is 1 million gallons per day (1.55 cubic feet per second); the seasonal pumping rate, shown by the magenta line, is approximately 4 million gallons per day (6.14 cubic feet per second) during June, July, and August. Depletion rates are shown for a well pumping at, *B,* 300 feet; *C,* 1,000 feet; and *D,* 3,000 feet from the river. Streamflow-depletion rates for the constant pumping rate are shown by the solid black lines and for the seasonal pumping rate by the magenta lines. The hydraulic diffusivity of the aquifer is 10,000 feet squared per day. [Rates of streamflow depletion were calculated by using a computer program described in Reeves (2008).]

The cyclic-pumping schedule results in cyclic patterns of streamflow depletion in the river, although the timing, rates, and range of depletion depend on the distance of the well from the river (fig. 21). The amplitude of the annual depletion rates is largest when the well is placed close to the river (that is, fig. 21B) but is substantially reduced as the distance of the well from the river is increased (fig. 21C and D). As noted by Jenkins (1968a), as the distance of the well from the river increases, the cyclic pumping pattern has an effect on stream-flow depletion that closely resembles the constant pumping pattern. This effect is illustrated in the figure by contrasting the patterns of streamflow depletion for pumping at distances of 300 and 3,000 ft from the river. For pumping at 300 ft, the annual range of depletion in the 15th year is 5.0 ft^3/s, whereas it is only 0.4 ft^3/s for pumping at a distance of 3,000 ft from the river. The figure also shows that for a constant rate of pumping at each well, streamflow-depletion rates asymptotically approach the pumping rate at each well (1.5 ft^3/s), although this constant rate of depletion is attained much more slowly as the distance of the well from the river is increased. In contrast, depletions that result from the cyclic-pumping schedules asymptotically approach a condition of annual dynamic equilibrium, and this condition is attained most slowly for pumping at a distance of 3,000 ft from the river.

The maximum rate of depletion for the well at 300 ft occurs on August 31 of each year, the last day of pumping, and the minimum depletion rate occurs on May 31, just before the well begins to pump for the new irrigation season. In contrast, for pumping at a distance of 3,000 ft, the maximum rate of depletion in the first year does not occur until December 21, more than 3 months after the irrigation period ends; the minimum depletion rate in the first year occurs on July 12 and 13, about half-way through the new irrigation season. In later years, the maximum rate of depletion for pumping at this well shifts to December 1—still 3 months into the non-irrigation season—and the minimum depletion shifts to July 18–19. The dependence of the timing of the maximum and minimum depletion rates on the distance of the well from the river has important implications to the management of streamflow depletion, which will be discussed later in the report.

For some time after the initiation of pumping, ground-water storage is the primary source of water to the well, and on an annual basis, the volume of depletion is less than the annual volume withdrawn (365 Mgal). With time, however, the annual volume of depletion approaches the annual volume pumped at the well, regardless of the distance of the well from the river or the pattern of withdrawal (constant or cyclic; Wallace and others, 1990; Darama, 2001; Bredehoeft, 2011a). The time required for the annual volume of depletion to equal the annual volume pumped increases with distance of the well from the river. In addition to the distance of the well from the river, the time required for the system to reach a new equilibrium is also a function of the hydraulic diffusivity of the aquifer and the width of the river valley (Butler and others, 2001; Miller and others, 2007; Bredehoeft, 2011a).

In summary, the effect of cyclic pumping close to a stream is highly variable depletion through time, with a maximum that may approach the maximum pumping rate during periods of pumping. In contrast, the effect of cyclic pumping at greater distances from a stream is less-variable depletion through time, with maximum depletion that may only be slightly greater than the long-term average pumping rate.

Multiple Wells and Basinwide Analyses

The focus of this report thus far has been on the effects on streamflow depletion caused by individual wells pumping at different locations within a groundwater system. Typically, however, multiple wells withdraw water simultaneously from locations distributed throughout a groundwater basin. Many groundwater basins in the United States have hundreds and in some cases thousands of wells from which water is withdrawn. Considered individually, these wells may have small effects on streamflow, but when evaluated together on the scale of an entire basin, these wells can have substantial effects on streamflow. Moreover, basinwide groundwater development typically occurs over a period of several decades, and the resulting cumulative effects on streamflow depletion may not be fully realized for many years. As a result of the large number of wells and complex history of development, it is often necessary to take a basinwide perspective to assess the effects of groundwater withdrawals on streamflow depletion.

Such an approach was taken in a study of the effects of groundwater development on streamflow in the Elkhorn and Loup River Basins of central Nebraska (fig. 22). Groundwater withdrawals from thousands of wells in these basins are used to irrigate crops, and the number of acres irrigated with groundwater has risen sharply since the 1940s (fig. 23; Peterson and others, 2008; Stanton and others, 2010). Withdrawals in the basins occur from the High Plains aquifer, with most of the wells located outside of the largely undeveloped Sand Hills region (Peterson and others, 2008). Total groundwater withdrawals within the areas shown in figure 22 averaged about 1,700 Mgal/d in 2005.

Groundwater pumping has had substantial effects on streamflow throughout the Elkhorn and Loup River Basins. These effects are illustrated by the cumulative reductions in groundwater discharge (base flow) to selected river reaches within the basins, as determined by use of a groundwater model of the area (fig. 24). Depletions were relatively small prior to 1970, but have increased sharply since then as the number of wells and total amount of withdrawals have

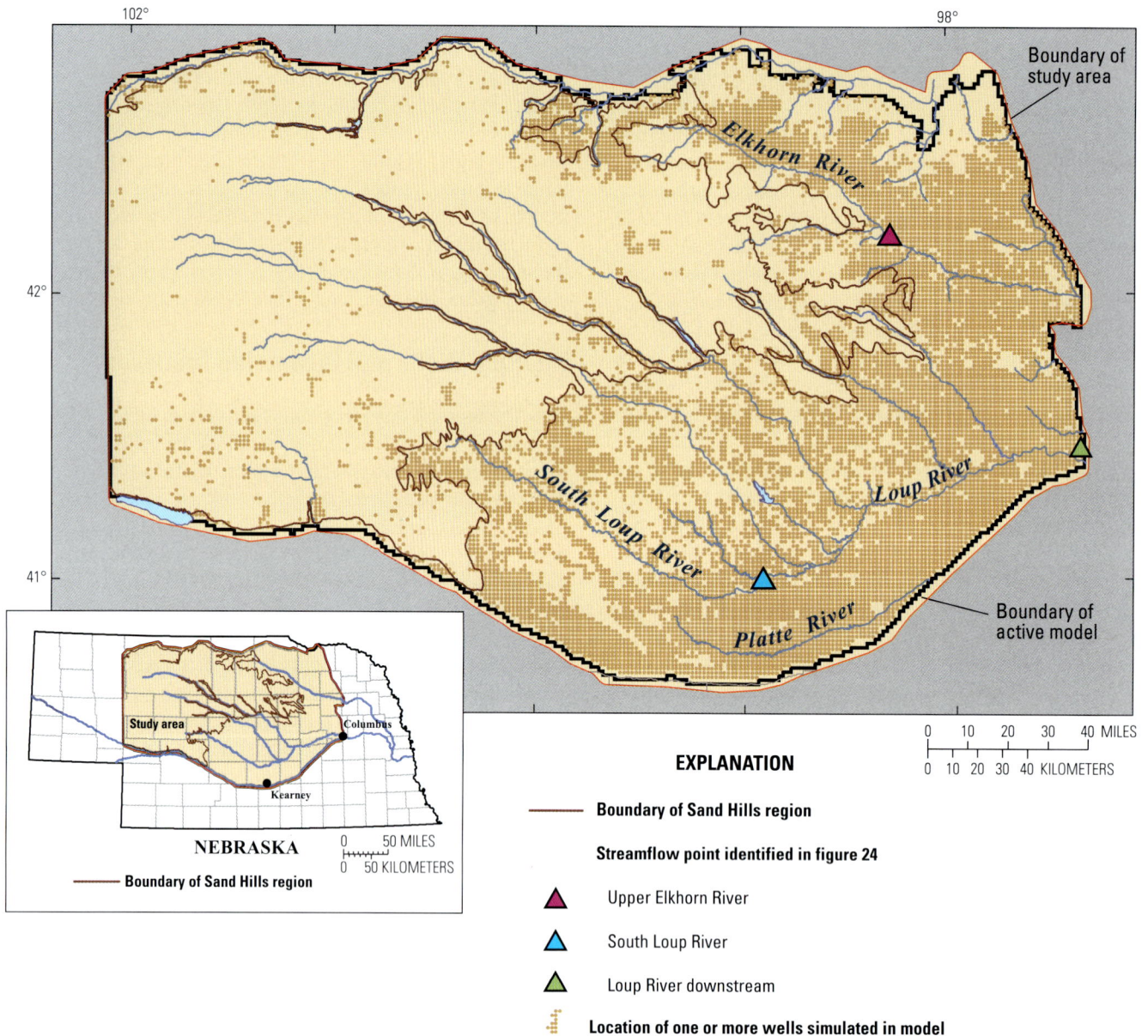

Figure 22. Locations of simulated pumping wells in parts of the Elkhorn and Loup River Basins, Nebraska, 2009. Locations of streamflow points identified in figure 24 are also shown (modified from Stanton and others, 2010).

increased. The effects of pumping have been largest for the lower reaches of the Loup River Basin, most likely because streams in those areas are in close proximity to extensive areas of irrigation (Peterson and others, 2008).

The basinwide effects of pumping in the Elkhorn and Loup River Basins over a period of several decades are representative of conditions that occur in groundwater basins throughout the United States. However, for the purpose of illustrating the underlying physical processes that occur when multiple wells pump from a groundwater system, it is useful to focus on just a few wells that withdraw water from a relatively simple aquifer system. As an example, the effects of a phased

groundwater-development program in which three wells are developed over a 15-year period are evaluated by use of the hypothetical system described in the previous section of the report, in which a single stream is bounded by an areally extensive aquifer. Development within the hypothetical system is assumed to progress over time from areas closest to the stream to those distant from the stream: well A, which is located 300 ft from the stream, begins pumping in year 1; well B, located 1,000 ft from the stream, begins pumping in year 6; and well C, located 3,000 ft from the stream, in year 11. Each well pumps at a constant rate of 1 Mgal/d (1.55 ft^3/s), such that the total rate of pumping is 1 Mgal/d for the first 5 years

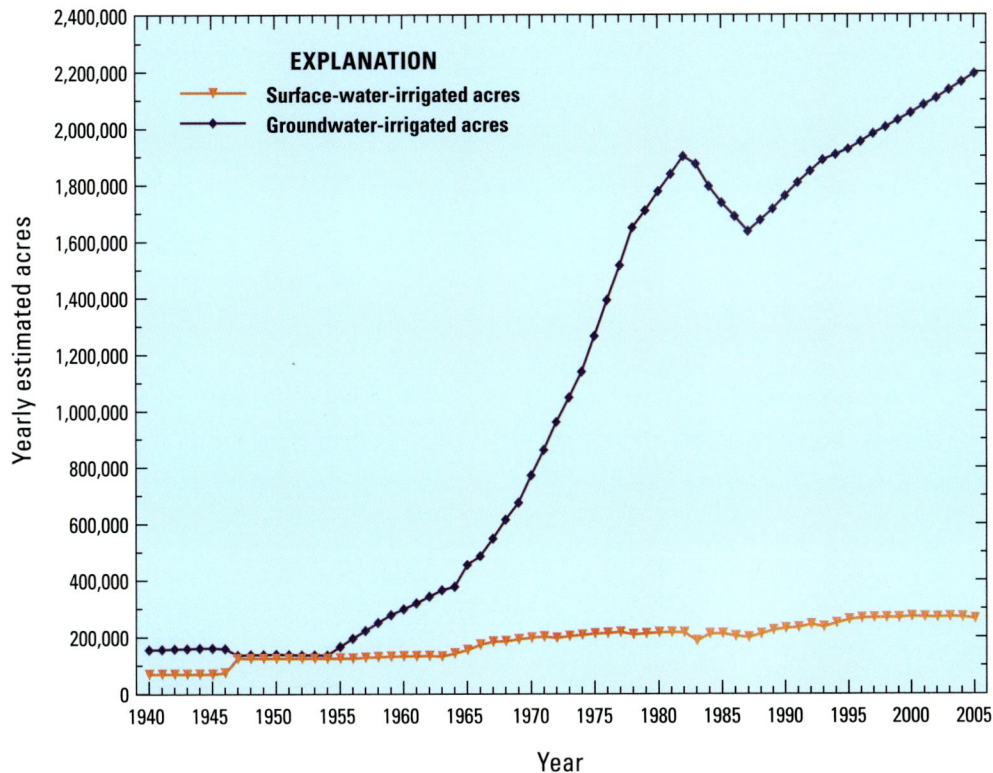

Figure 23. Estimates of acres of cropland irrigated by groundwater and surface water, 1940 through 2005, Elkhorn and Loup River Basins, Nebraska (modified from Stanton and others, 2010).

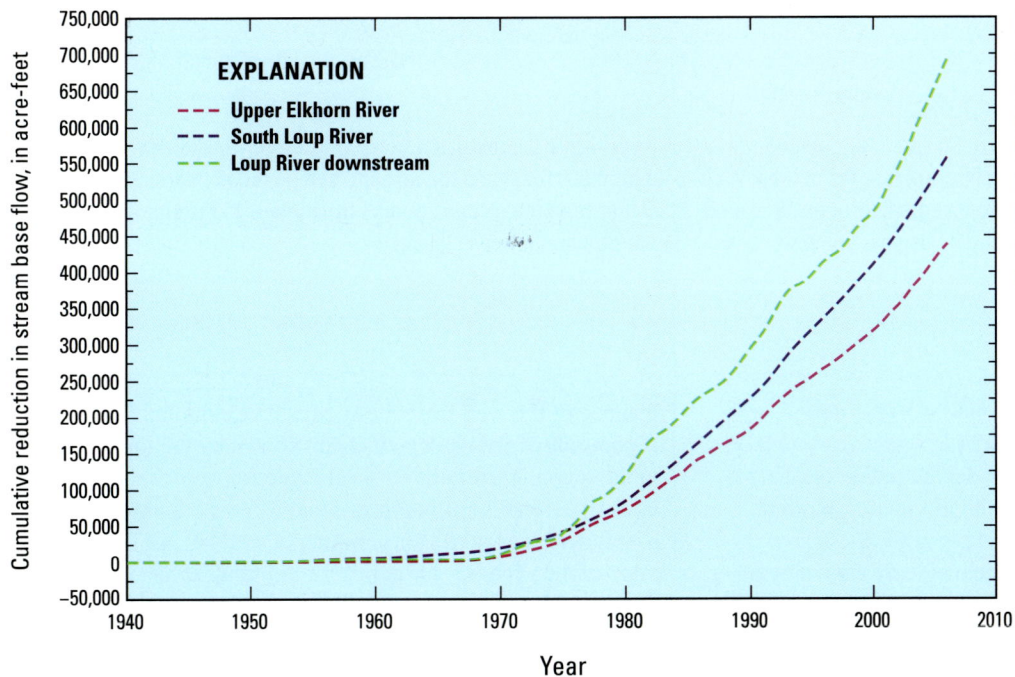

Figure 24. Model-calculated cumulative reductions in stream base flow caused by groundwater pumping, Elkhorn and Loup River Basins, Nebraska, 1940 through 2005. Locations of streamflow points shown in figure 22 (modified from Peterson and others, 2008).

A

B

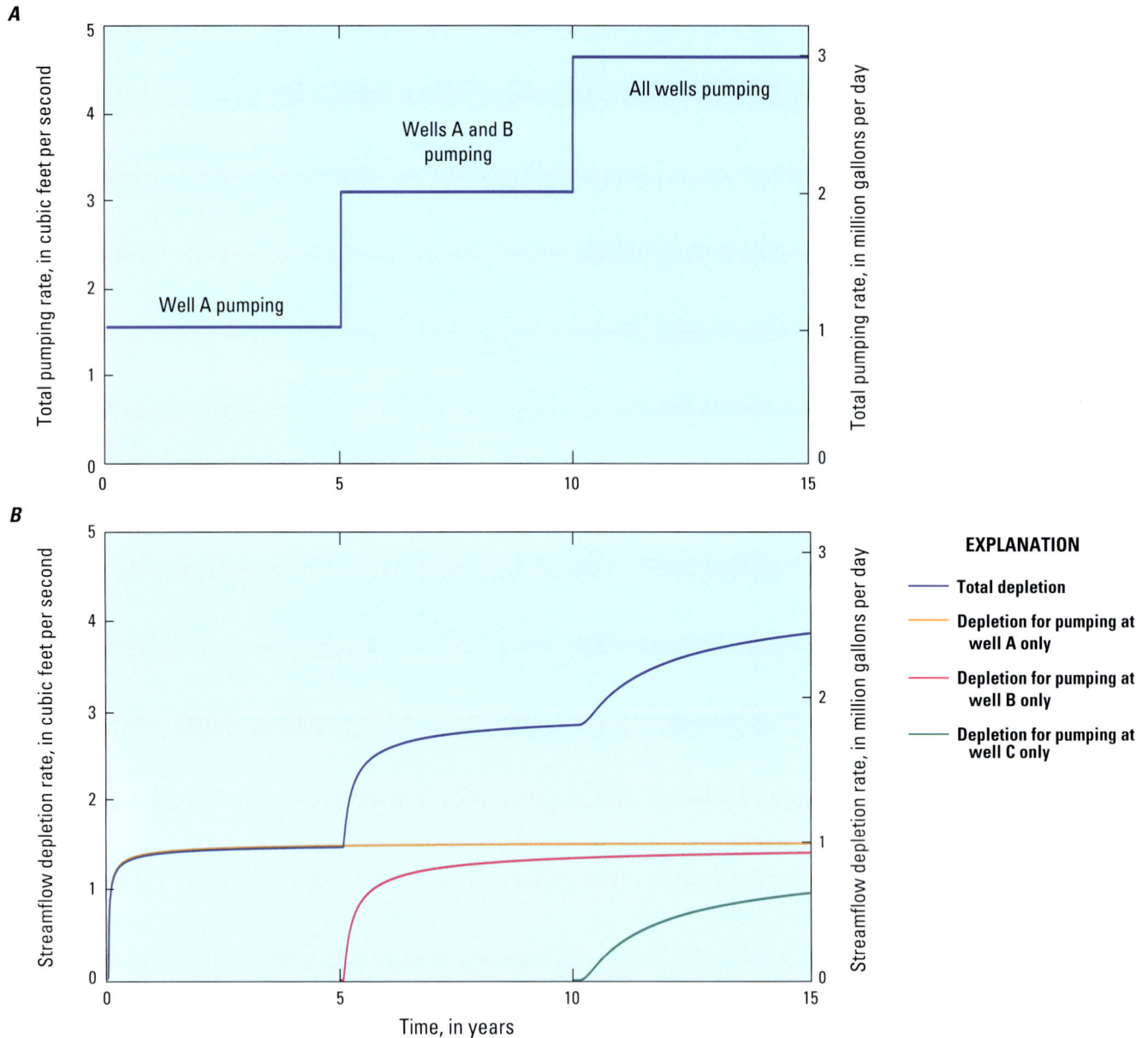

Figure 25. *A,* Total pumping rates. *B,* Streamflow-depletion rates for three wells pumping at a constant rate of 1 million gallons per day (1.55 cubic feet per second) for different lengths of time. Well A, located 300 feet from the stream, pumps for 15 years; well B, 1,000 feet from the stream, pumps from years 6 through 15; well C, 3,000 feet from the stream, pumps from years 11 through 15. [Rates of streamflow depletion were calculated by using a computer program described in Reeves (2008).]

of development, 2 Mgal/d for the middle 5-year period, and 3 Mgal/d for the last 5 years (fig. 25*A*).

Figure 25*B* shows streamflow depletion that occurs in response to pumping at each well individually, as well as the cumulative effects of pumping at all wells. The individual effects of pumping at each well are shown by the lower three curves in the figure, which indicate that streamflow depletion asymptotically approaches the pumping rate of each well, 1.55 ft^3/s (1.0 Mgal/d), regardless of the

distance of the well from the stream. The cumulative effects of pumping at the wells, which are shown by the top curve on the graph, are additive—total depletion approaches a value of 1.55 ft^3/s for pumping at well A only, a value of 3.10 ft^3/s (2.0 Mgal/d) for pumping at wells A and B, and a value of 4.65 ft^3/s (3.0 Mgal/d) for pumping at all three wells, although the time at which this depletion rate would be fully realized occurs several years after the 15-year time frame evaluated here.

Pumped Wells and Recharge Wells

The practice of artificial recharge of water into aquifers is becoming an increasingly important component of many water-resource management programs. Artificial recharge is used as an alternative to surface-water reservoirs to store excess surface water and as a means to augment stream-flows. Methods for artificially recharging an aquifer include direct injection by wells and infiltration by gravity in basins or ponds at the land surface. When water is injected into an aquifer at a recharge well, groundwater levels near the well increase, and groundwater flows outward from the resulting area of mounded water. If the aquifer is bounded by a stream, the rate of groundwater discharge to the stream will increase, and the timing and rate of streamflow accretion will be equal, but opposite in sign, to the timing and rate of streamflow depletion caused by pumping at the same location and rate (as long as the system responds linearly to pumping, which is discussed in the section on "Superposition Models"). This scenario is illustrated by the first two

Photograph by Michael Collier

Braided channel of the Platte River, Nebraska.

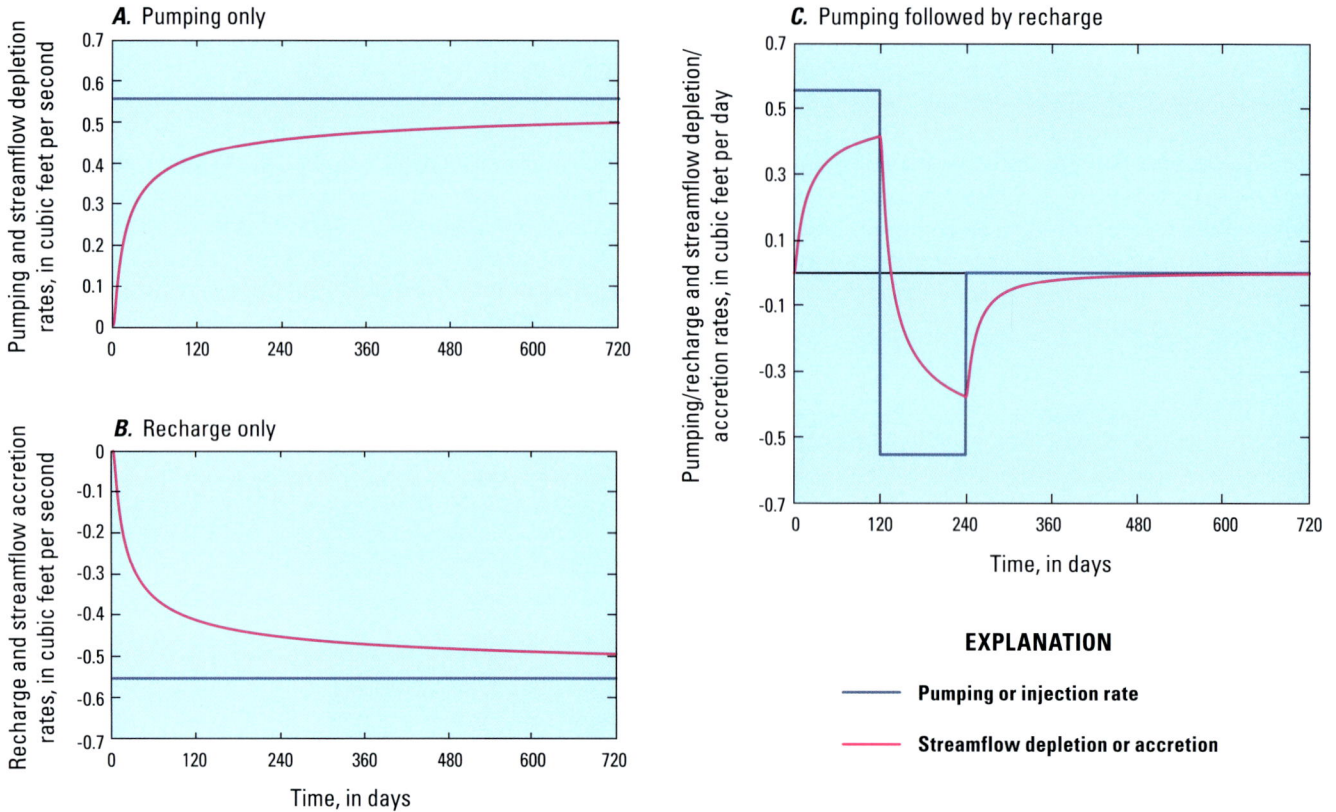

Figure 26.   Injection of water into a recharge well increases streamflow, and the timing and rates of streamflow accretion are equal, but opposite in sign, to those of streamflow depletion caused by pumping. *A*, A well located 500 feet from a stream is pumped at a rate of 250 gallons per minute (0.56 cubic foot per second) for 720 days. *B*, The same well is recharged at a rate of 250 gallons per minute for 720 days. *C*, The well is pumped for 120 days, followed by a 120-day period of recharge. [Rates of streamflow depletion and accretion were calculated by using a computer program described in Reeves (2008).]

graphs in figure 26, which contrast streamflow depletion caused by pumping at a well located 500 ft from a stream at a rate of 250 gallons per minute (gal/min; fig. 26*A*) with streamflow accretion caused by recharging the aquifer at the same well at a rate of 250 gal/min (fig. 26*B*). The shape of the streamflow-depletion and streamflow-accretion curves are mirror images of one another, and each curve tends asymptotically toward the pumping or recharge rate of the well (±0.56 ft³/s or ±250 gal/min). Because depletion has been represented as a positive quantity throughout this report, streamflow accretion is shown as a negative quantity in the figure, although it should be apparent that artificial recharge has a positive effect on streamflow. The results shown in the figure are based on the important assumption that the mound

of groundwater that is formed by injection at the recharge well remains below land surface; should the mound reach land surface, surface-water runoff may occur, resulting in less water available for groundwater discharge to the stream.

Graph *C* in figure 26 illustrates the effects of a 120-day pumping period followed by a 120-day recharge period at the same well. As described in the previous section of the report for the case of multiple wells pumping from an aquifer, the combined effects of the pumping and recharge cycle on streamflow are additive. As a result, the period of streamflow depletion caused by pumping is followed by a period of streamflow accretion caused by recharge; ultimately, sometime after recharge ends, the effects of pumping and recharge at the well diminish to zero.

Streamflow Depletion and Water Quality

One of the important concerns associated with streamflow depletion by wells is the effect of reduced groundwater discharge on the quality of affected surface waters. Groundwater discharge affects the chemistry of surface water and plays an important role in regulating stream temperature, which is a critical water-quality property in determining the overall health of an aquatic ecosystem (Baron and others, 2002; Hayashi and Rosenberry, 2002; Stonestrom and Constantz, 2003; Risley and others, 2010). Because groundwater-temperature fluctuations are relatively small compared to daily and seasonal streamflow-temperature fluctuations, groundwater discharge at a nearly constant temperature provides a stable-temperature environment for fish and other aquatic organisms. Average shallow groundwater temperature at a particular location is approximately equal to mean annual air temperature, and, as a result, groundwater discharge is typically warmer than the receiving streamflow during the winter and cooler than the receiving streamflow during the summer. Groundwater discharge provides cool-water environments that protect fish from excessively warm stream temperatures during the summer, and conversely, relatively warm groundwater discharge can protect against freezing of the water during the winter (Hayashi and Rosenberry, 2002). Stark and others (1994) and Risley and others (2010) provide examples of the effects of pumping on stream temperatures. The work of Risley and others (2010) illustrates how reductions in the rates of groundwater discharge to streams caused by pumping can warm stream temperatures during the summer and cool stream temperatures during the winter.

For many issues related to the quantity of streamflow depletion, such as water-rights administration and instream-flow needs to sustain aquatic habitats, the distinction between the two components of depletion—captured groundwater discharge and induced infiltration of streamflow—is generally not of interest. For water-quality concerns, however, the relative contribution of captured groundwater discharge and induced infiltration have important implications to the resulting quality of the streamflow, groundwater, and pumped water. Where groundwater pumping is large enough to cause induced infiltration of streamflow, the quality of the induced surface water will affect the quality of water in the underlying aquifer and possibly that of the pumped wells themselves. Infiltrated surface water that has been contaminated by chemical pollutants or biological constituents such as *Giardia*

lamblia and *Cryptosporidium,* therefore, can be a source of contamination to a groundwater system, potentially having adverse effects to the health of people ingesting water from the contaminated groundwater supply. The amount of surface-water contamination entering a water-supply well will depend on several factors, including the natural ability of the streambank and aquifer materials to filter contaminants from the polluted water (Bourg and Bertin, 1993; Macler, 1995). Natural "bank filtration" of surface-water contaminants as they move from a stream to a pumped well involves geochemical and biological processes that remove nutrients, organic carbon, and microbes from the contaminated water (National Research Council, 2008; Farnsworth and Hering, 2011). Numerous field studies of the distribution, transport, and fate of chemical and biological constituents within contaminated and uncontaminated aquifers have been done to establish hydraulic connections between surface-water sources and pumped groundwater and to test the effectiveness of bank filtration and other natural processes for reducing contaminant concentrations. Examples of these types of studies are available for many areas of Europe and the United States (Farnsworth and Hering, 2011), including Ohio (Sheets and others, 2002), Missouri (Kelly, 2002; Kelly and Rydlund, 2006), and Oregon (McCarthy and others, 1992).

Groundwater-temperature measurements can be an effective method to demonstrate the hydraulic connection that exists between groundwater and surface-water systems and to trace surface-water infiltration in groundwater systems (Stonestrom and Constantz, 2003; Anderson, 2005; Constantz, 2008). An example of the use of temperature measurements to demonstrate a hydraulic connection between surface water and pumped wells is provided by the results of a study conducted in 1960–61 along the Mohawk River near Schenectady, New York (fig. 27), where the aquifer consists of highly permeable sand and gravel deposits. Groundwater pumped from two well fields near the river averaged about 20 Mgal/d during the study period, with about 90 percent of the pumping occurring from the well field furthest from the river (Winslow, 1962). The temperature of the river on the measurement date (September 7, 1961) was 77 degrees Fahrenheit (°F), which was nearly 30 °F warmer than the average temperature of the groundwater in areas unaffected by induced infiltration. The warm river water, which was drawn into the aquifer by pumping at the production wells, became progressively cooler with distance from the river as it mixed with the cold groundwater.

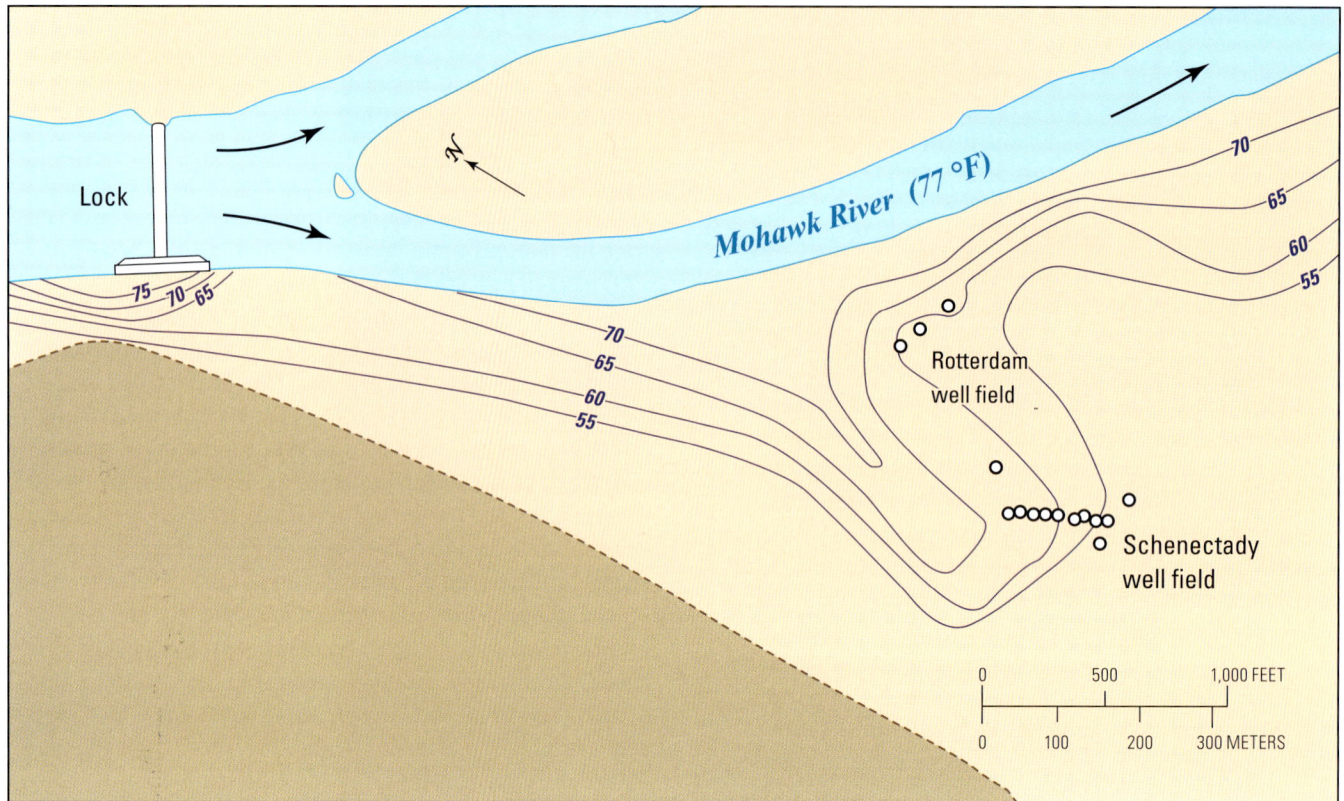

Figure 27.   Groundwater-temperature contours in the vicinity of two well fields near the Mohawk River, New York, on September 7, 1961. Contours are based on measurements of groundwater temperature made at 60 observation wells. Temperature of the river on that date was 77 degrees Fahrenheit (modified from Winslow, 1962).

Some of the factors that affect the relative contributions of captured groundwater discharge and induced infiltration can be illustrated by one of the hypothetical stream-aquifer systems described previously and shown in figure 14. Several steady-state and transient simulations were done with the numerical model of the stream-aquifer system for pumping at a rate of 1.0 Mgal/d at wells located 100 ft, 300 ft, 700 ft, and 1,400 ft from the stream. Steady-state conditions were simulated to illustrate the maximum effects of pumping on streamflow. The long-term average recharge rate to the aquifer of 26.0 inches per year (in/yr) also was varied in these simulations to include a 25-percent increase in the recharge rate (32.5 in/yr) and a 25-percent decrease in the recharge rate (19.5 in/yr). For each simulation, the resulting rates of total streamflow depletion, captured groundwater discharge, and induced infiltration were determined at the outflow point of

the basin (that is, the most downstream location on the stream in figure 14).

Results for the steady-state simulations are shown in figure 28. As shown by the uppermost curve in the figure, the total amount of streamflow depletion at the outflow point of the basin is the same for all of the simulations (and equal to the 1.0 Mgal/d pumping rate at each of the wells), regardless of either the distance of the well from the stream or the recharge rate to the aquifer. This results from the fact that at equilibrium, when aquifer storage is no longer a source of water to the wells, all of the water pumped by the wells must result in decreased streamflow, either by captured groundwater discharge or by induced infiltration. In contrast, the relative contributions of captured discharge and induced infiltration are a function of both the distance of the well from the stream and the recharge rate to the aquifer. As the well distance

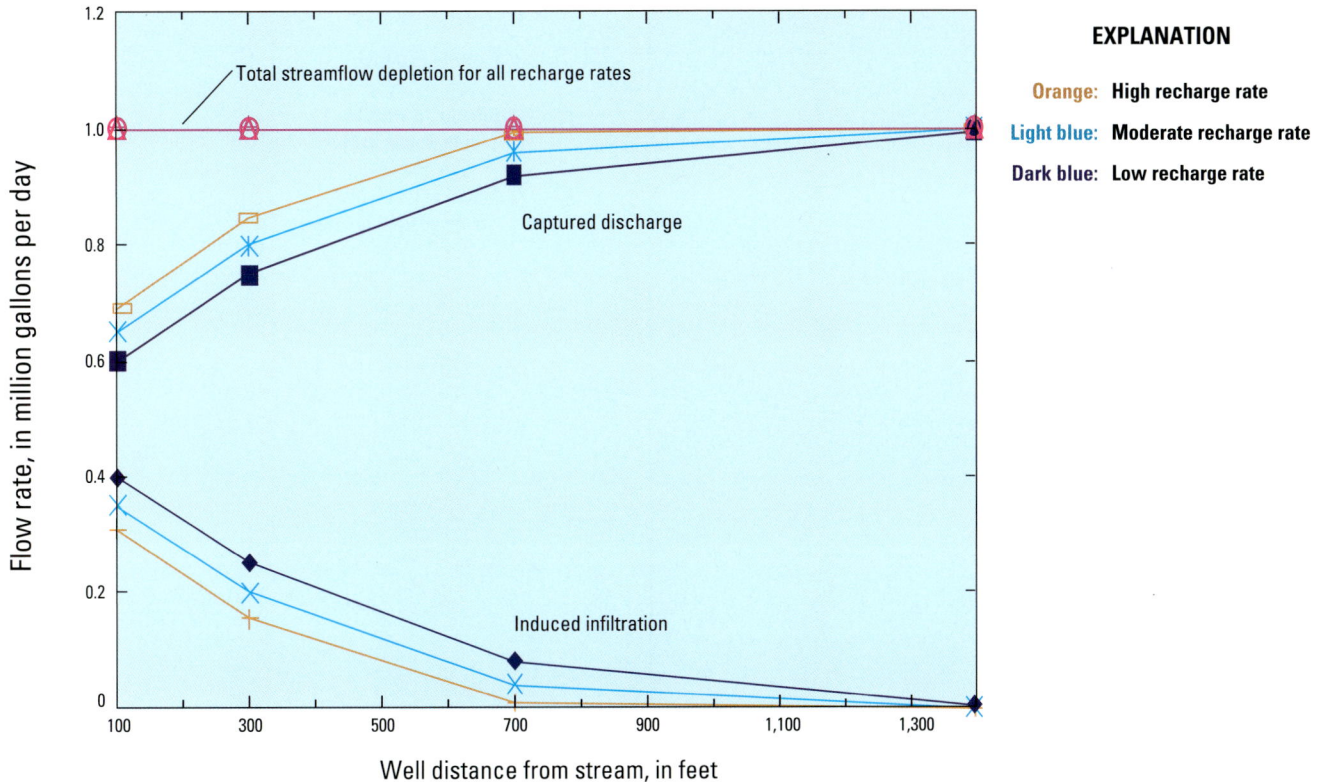

Figure 28. Rates of streamflow depletion, captured groundwater discharge, and induced infiltration at the outflow point of a basin for steady-state pumping conditions at wells located 100, 300, 700, and 1,400 feet from a stream. Each well is pumped independently of the others at a rate of 1.0 million gallons per day in 12 separate simulations. Three rates of recharge were simulated: a high recharge rate (32.5 inches per year), a moderate recharge rate (26.0 inches per year), and a low recharge rate (19.5 inches per year). Total streamflow depletion is equal to the sum of captured discharge and induced infiltration. As shown by the results along the top (pink) curve, at steady state, the total rate of streamflow depletion at the basin outflow point is equal to the pumping rate of each well and is independent of both the distance of each well from the stream and the recharge rate to the aquifer. Rates of captured groundwater discharge (middle three curves) and induced infiltration (bottom three curves), however, are a function of both well distance from the stream and recharge rate to the aquifer. (Results from numerical model shown in figure 14 of this report and documented in Barlow, 1997.)

from the stream increases, the proportion of induced infiltration decreases. Similarly, as the recharge rate to the aquifer increases, the proportion of induced infiltration also decreases. Note that at a well distance of 1,400 ft, the rate of induced infiltration for all recharge rates is essentially zero. These results illustrate that for some stream-aquifer systems, such as the extensive systems found in the Midwestern and Western United States in which pumping occurs miles from a stream, induced infiltration may not occur, and the water-quality concerns associated with streamflow depletion will be focused on the effects of reduced groundwater discharge on the thermal and water-quality conditions of the receiving streams. The results of additional simulations for transient-flow conditions were consistent with the steady-state simulations; specifically, the rate of recharge affects the relative contributions of

captured discharge and induced infiltration but does not affect the total rate of streamflow depletion.

Other factors also affect the relative proportion of captured discharge and induced infiltration. These factors include the pumping rate of the well (with greater rates of induced infiltration occurring for higher pumping rates), the direction of groundwater flow in the aquifer prior to pumping, the distribution of aquifer boundaries near the well (including the presence of impermeable boundaries and other streams), the hydraulic properties of the aquifer and streambank materials, and the penetration depths of the pumped well and stream into the aquifer (Newsom and Wilson, 1988; Morrissey, 1989; Wilson, 1993; Conrad and Beljin, 1996; Chen, 2001; Chen and Yin, 2001; Chen and Shu, 2002; Chen and Chen, 2003; Chen and Yin, 2004; Gannett and Lite, 2004).

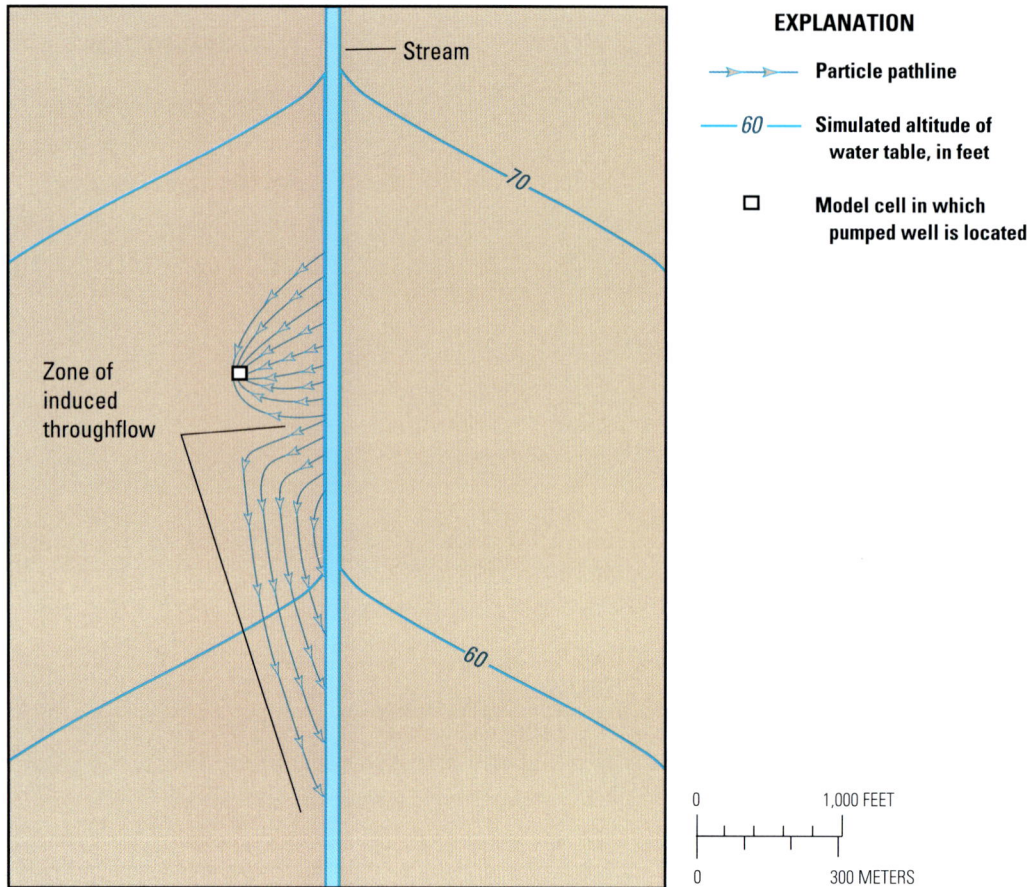

Figure 29. Pumping at the well located 300 feet from the stream at a rate of 1.0 million gallons per day causes induced infiltration of streamflow. More than 80 percent of the induced streamflow is captured by the well, but some of the induced streamflow returns to the stream through a zone of "induced throughflow" (Newsom and Wilson, 1988; model results from Barlow, 1997).

Several of these factors also affect the proportion of induced infiltration that actually flows to and is discharged by a well. Figure 29 shows the flow paths of water particles that have been drawn into the aquifer by pumping at the well 300 ft from the stream at the steady-state rate of 1.0 Mgal/d (for the average recharge rate of 26.0 in/yr). As shown by the flow paths, some of the water that has been drawn into the aquifer actually returns to the stream downgradient from the well and does not reach the well. Newsom and Wilson (1988) refer to the area in which induced infiltration flows back to the stream as the "zone of induced throughflow." The figure illustrates that the rate of stream infiltration is not the same as the rate of infiltrated streamflow that is actually pumped at the well. The ability to quantify the relative contributions of captured discharge and induced infiltration to the source of water pumped by a well, or the concentrations of chemical constituents in the well or adjoining aquifer, requires analysis techniques that are more advanced than those used to quantify streamflow-depletion rates only. These techniques include computer programs that track water particles through a simulated aquifer, such as illustrated by the flow paths shown in figure 29 that were calculated by use of MOD-PATH (Pollock, 1994), or solute-transport codes that simulate movement of chemical constituents within a groundwater-flow system (for example, the computer programs documented by Konikow and others, 1996, or Zheng and Wang, 1999).

Common Misconceptions about Streamflow Depletion

An understanding of the basic concepts of streamflow depletion is needed to properly assess the effects of groundwater withdrawals on connected surface water and areas of evapotranspiration. Important concepts relating to depletion are available throughout this report and also in other literature, beginning with the paper, "The Source of Water Derived from Wells," by Theis (1940). In spite of these sources of information, misconceptions regarding factors controlling depletion are sometimes evident in analyses of depletion. This discussion highlights the following common misconceptions related to streamflow depletion.

Misconception 1. Total development of groundwater resources from an aquifer system is "safe" or "sustainable" at rates up to the average rate of recharge.

Misconception 2. Depletion is dependent on the rate and direction of water movement in the aquifer.

Misconception 3. Depletion stops when pumping ceases.

Misconception 4. Pumping groundwater exclusively below a confining layer will eliminate the possibility of depletion of surface water connected to the overlying groundwater system.

Although most of the concepts needed to clear up these misconceptions are presented in other sections, further discussion and examples are given here.

Aquifer Recharge and Development of Water Resources

There has been a tendency in parts of the United States to view groundwater development in an aquifer to be "sustainable" or "safe" when the overall rate of groundwater extraction does not exceed the long-term average rate of recharge to the aquifer. Conversely, development is considered to be unsustainable or unsafe when groundwater extraction rates exceed the average recharge rate. The rationale behind this concept is that long-term extraction beyond the average recharge rate will result in ongoing net depletions in storage that will eventually deplete the aquifer to the extent that continued pumping is no longer feasible. These views of sustainability, however, do not directly recognize the effects of withdrawals on outflow from an aquifer, which often occur through groundwater discharge to surface-water features and through evapotranspiration.

In the paper "Groundwater—The Water-Budget Myth," Bredehoeft and others (1982) explained that in an undeveloped aquifer, long-term average natural recharge is balanced by long-term average natural discharge. They show that if water is pumped from the aquifer at a given rate, that rate will be offset by the sum of an increase in the rate of recharge to the aquifer, decrease in the rate of discharge from the aquifer, and increase in the rate of removal of water from storage in the aquifer. With time, the rate of removal of water from storage change diminishes and the pumping rate is balanced by the sum of pumping-induced increased recharge and decreased discharge.

Most recharge to aquifers occurs through percolation of a portion of precipitation from the land surface, through an unsaturated zone, to the water table. In more humid areas, this recharge can be widely distributed over the surface area of an aquifer, and in more arid areas, this recharge can be focused in locations such as beneath arroyos where infrequent runoff events cause a movement of water through the unsaturated zone. In either case, however, the process of natural recharge through the unsaturated zone is unaffected by a pumped well. On the contrary, one situation in which pumping can increase recharge occurs in areas in which the water table is at the land surface (fig. 30*A*). Drawdown from pumping can result in infiltration and recharge that would have otherwise run off because of a lack of available space for storage beneath the land surface (fig. 30*B*). Another situation in which pumping can increase recharge is when recharge occurs from direct movement of water from surface-water bodies to the aquifer, such as for a naturally losing stream; this type of increased recharge is a form of induced infiltration.

Discharge from aquifers, on the other hand, commonly occurs through direct movement of groundwater into surface-water bodies and through evaporation and transpiration by plants that use groundwater. Groundwater pumping reduces the movement of water into surface-water features by decreasing the natural hydraulic gradients to these features. Pumping furthermore reduces evaporation and transpiration by lowering the water table below the land surface and roots of plants that use groundwater.

In spite of several possible cases in which pumping can increase recharge to an aquifer, most recharge is unaffected by pumping. Therefore, increases in recharge from pumping often can be considered to be small or zero. In this case, the pumping rate eventually will be approximately balanced by decreases in discharge. For this reason, Bredehoeft and others (1982) concluded that the magnitude of sustained groundwater pumping generally depends only on how much of the natural discharge can be captured. Although there may be physical limits to the amount of water that can be captured, lower limits to capture may exist for other reasons. For example, certain levels of instream flow may be required to sustain aquatic ecosystems, and capture or depletion of surface water that diminishes flow below those limits may not be permitted under some regulatory systems. Similarly, depletion that reduces the availability of surface-water flow for holders of surface-water rights may not be permitted in some areas. For further discussions of sustainability of groundwater resources, see Alley and others (1999), Alley and Leake (2004), and Gleeson and others (2012).

A

B

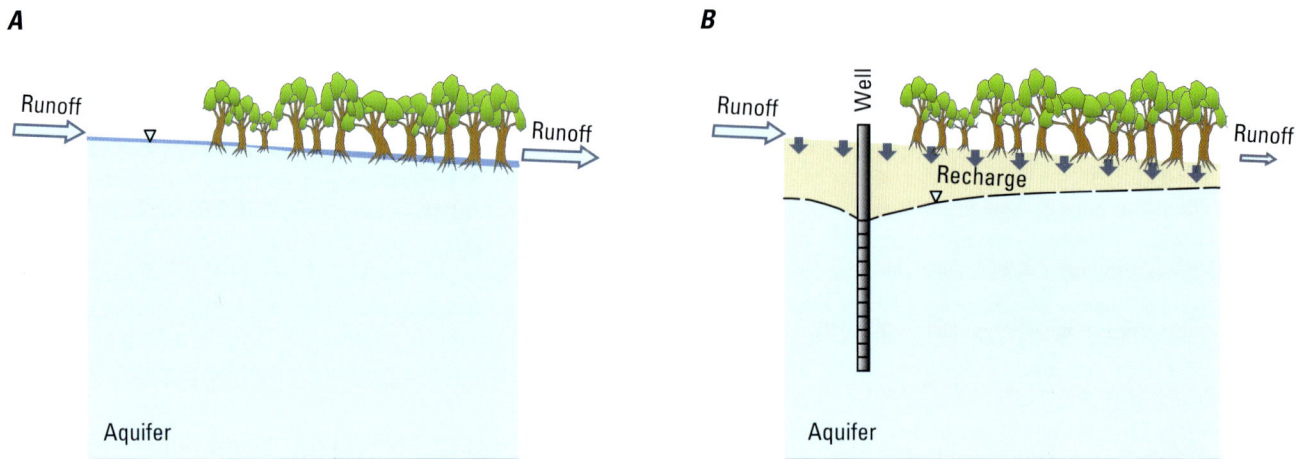

Figure 30.   One way in which groundwater pumping can increase recharge to an aquifer. *A,* If the water table is at land surface, surface runoff cannot infiltrate because of a lack of available subsurface storage space. *B,* The addition of a pumped well that lowers the water table allows runoff to infiltrate and recharge the aquifer.

Depletion and the Rates and Directions of Groundwater Flow

A common misunderstanding regarding streamflow depletion is that the rates, locations, and timing of depletion are dependent on the pre-pumping rates and directions of groundwater flow in an aquifer. As indicated previously, depletion is the sum of pumping-induced increased inflow to the aquifer and decreased outflow from the aquifer. Provided that sufficient surface water is available to meet the pumping demand, a new steady-state condition will eventually be reached in which the rate of storage change is zero and the entire pumping rate can be accounted for as increased recharge and decreased discharge. It is important to understand that depletion is independent of the natural, pre-pumping rates of recharge and discharge. The concept that the rate of recharge does not affect the rate of depletion was demonstrated previously (fig. 28), where it was also noted that the recharge rate does affect the individual components of depletion— captured groundwater discharge and induced infiltration. Maddock and Vionnet (1998) extended these concepts to show that even with seasonally varying recharge and discharge, temporal patterns of recharge and discharge do not enter into calculations of depletion.

Timing and locations of depletion are affected, however, by aquifer properties and system geometry. In a system with predominantly horizontal flow, the progression from a storage-dominated to a depletion-dominated supply of pumped water is controlled by hydraulic diffusivity (Box A) and distance between pumping locations and connected surface-water and groundwater-evapotranspiration areas or other groundwater-discharge areas. In settings in which vertical components of groundwater flow are important, distributions of vertical and horizontal hydraulic conductivity, specific storage, specific yield, and aquifer thickness, in addition to well distance from the stream, are the key properties that control the timing of depletion.

As long as aquifer transmissivity and storage properties are the same in each case, total depletion (in contrast to the individual components of captured discharge and induced infiltration) at any given time would be the same for cases with natural pre-pumping flow from the stream to the aquifer (fig. 31*A*), from the aquifer to the stream (fig. 31*B*), or with no flow between the aquifer and the stream (fig. 31*C*). Furthermore, relative amounts of depletion in multiple streams are the same regardless of the existence of a divide between the streams (fig. 31*D*), natural flow from one stream to another (fig. 31*E*), or with no flow between the streams (fig. 31*F*).

The independence of depletion and rates and directions of groundwater flow in most systems allows calculation of depletion by a number of different methods. These methods include analytical solutions, superposition models, and groundwater-flow models (see "Analytical and Numerical Modeling" section). In using either analytical solutions or superposition models, the natural rates and directions of groundwater flow are ignored.

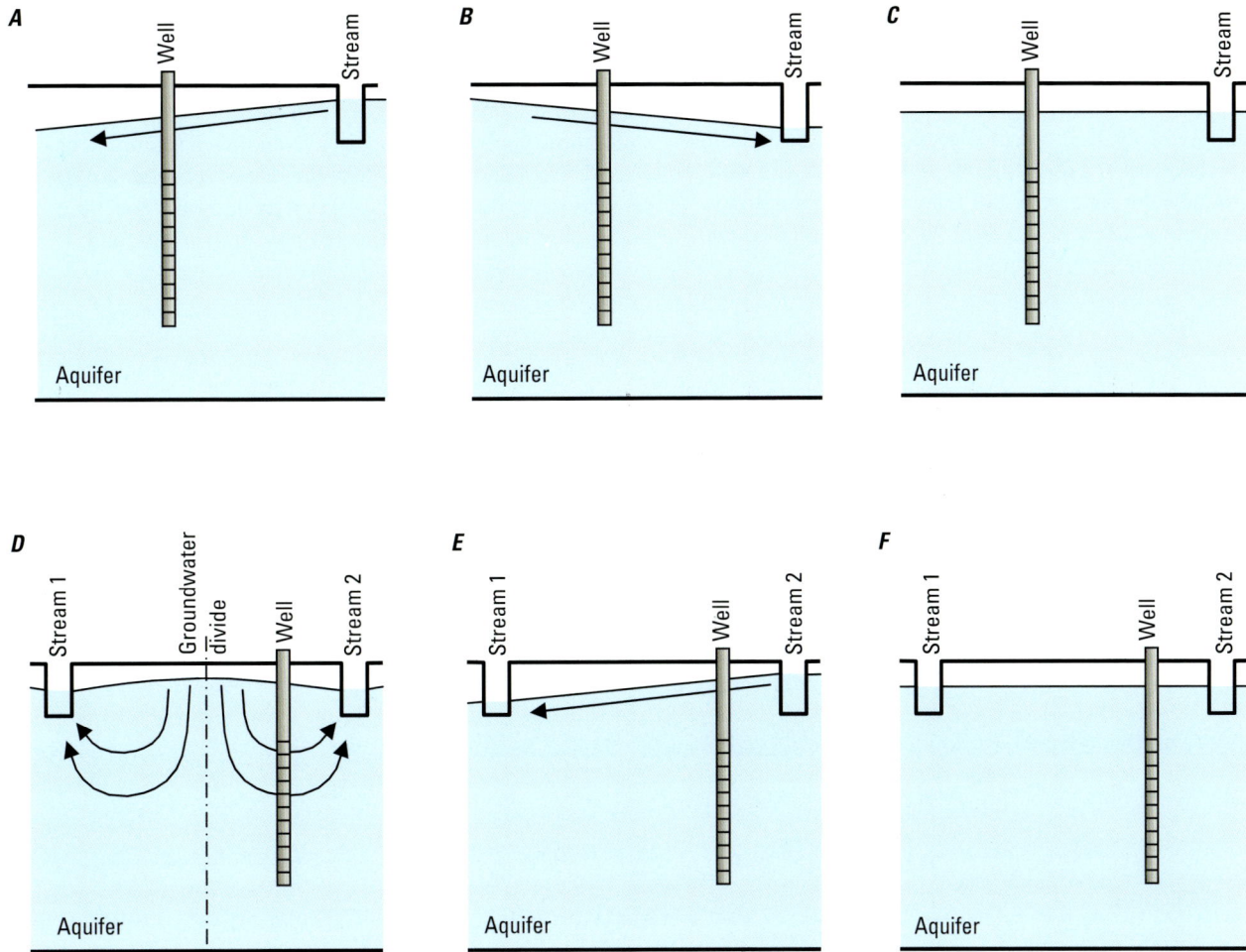

EXPLANATION

⟵———————   Direction of groundwater flow
before pumping

Figure 31. Position of a pumped well in relation to a stream or streams for configurations of various pre-pumping groundwater-flow patterns. As long as aquifer properties are the same in each case, depletion of the steam by the pumping well would be the same with, *A*, pre-pumping flow away from the stream; *B*, pre-pumping flow toward the stream; or, *C*, no pre-pumping flow. Similarly, relative amounts of depletion in adjacent streams are unaffected by a groundwater divide with, *D*, pre-pumping flow toward each stream; *E*, pre-pumping flow from one stream to the other; or, *F*, no flow between streams.

Depletion after Pumping Stops

When a well begins to pump, water is removed from storage around the well, creating a cone of depression. As discussed previously, the cone of depression expands and can increase recharge to and discharge from the aquifer. If a well pumps groundwater for a period of time and then pumping ceases, groundwater levels will begin to recover and the cone of depression created by the pumping will gradually fill, with water levels eventually reaching positions that existed before pumping started (fig. 32). During the time that the cone of depression is filling, groundwater that otherwise would have flowed to streams instead goes into aquifer storage; thus, streamflow depletion is ongoing, even though pumping has ceased. The factors that control the rate of recovery are the same as those that affect the rate of groundwater-level declines in response to pumping—the geology, dimensions, and hydraulic properties of the groundwater system; the locations

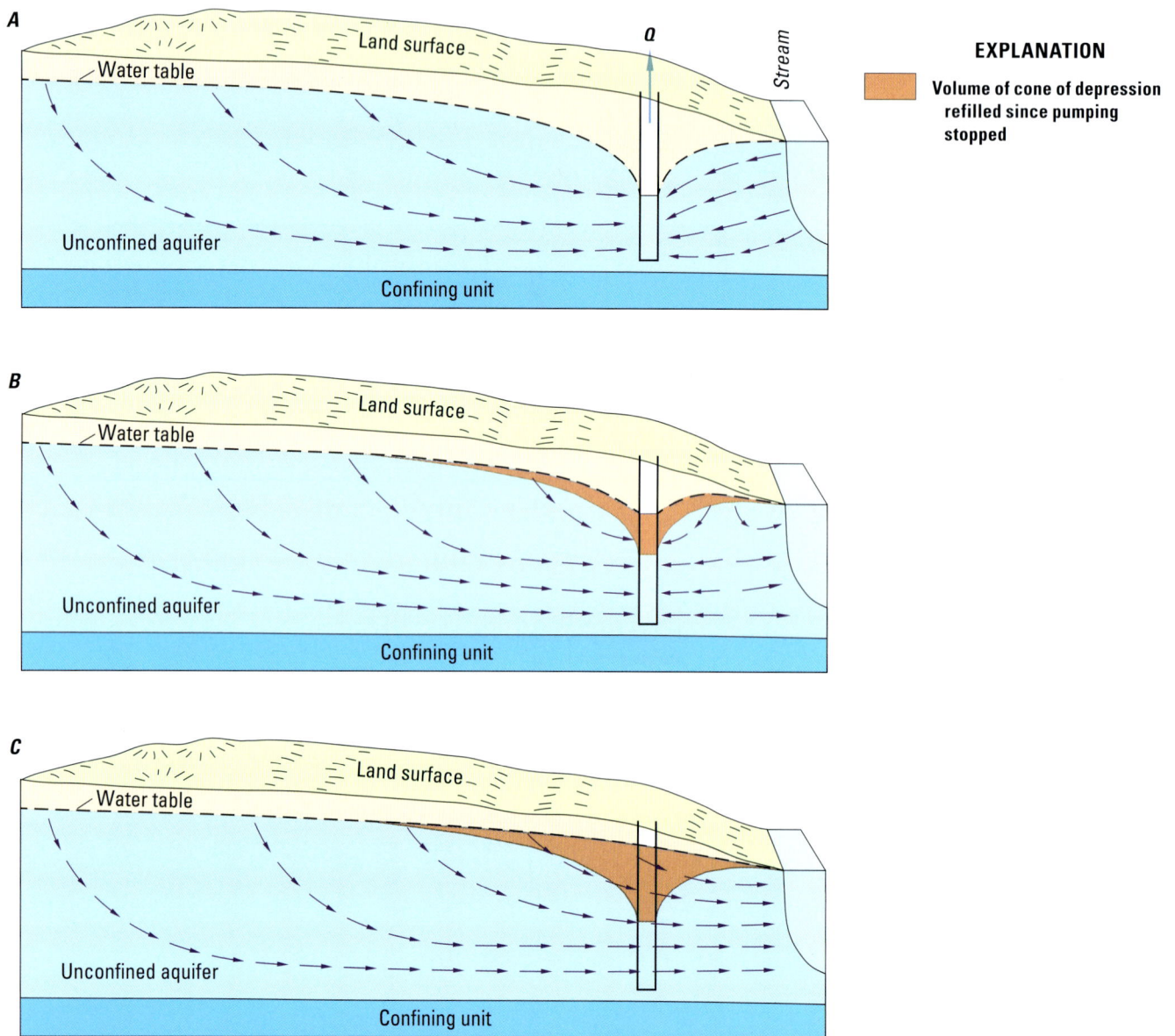

Figure 32.   Residual effects of streamflow depletion after pumping stops. *A,* Prior to the well being shut down, the pumping rate at the well is balanced by decreases in aquifer storage and by streamflow depletion, which consists of captured groundwater discharge and induced infiltration of streamflow. *B,* After pumping stops, groundwater levels begin to recover, and water flows into aquifer storage to refill the cone of depression created by the previous pumping stress. *C,* Eventually, the system may return to its pre-pumping condition with no additional changes in aquifer storage or streamflow depletion. [*Q,* pumping rate at well]

and hydrologic conditions along the boundaries of the groundwater system, including the streams; and the horizontal and vertical distance of the well from the stream.

Some key points relating to depletion from a well or wells that pump and then stop pumping are as follows:

1. Maximum depletion can occur after pumping stops, particularly for aquifers with low diffusivity or for large distances between pumping locations and the stream.

2. Over the time interval from when pumping starts until the water table recovers to original pre-pumping levels, the volume of depletion will equal the volume pumped.

3. Higher aquifer diffusivity and smaller distances between the pumping location and the stream increase the maximum rate of depletion that occurs through time, but decrease the time interval until water levels are fully recovered after pumping stops.

4. Lower aquifer diffusivity and larger distances between the pumping location and the stream decrease the maximum rate of depletion that occurs through time, but increase the time interval until water levels are fully recovered after pumping stops.

5. Low-permeability streambed sediments, such as those illustrated in figure 11, can extend the period of time during which depletion occurs after pumping stops.

6. In many cases, the time from cessation of pumping until full recovery can be longer than the time that the well was pumped.

Most of these concepts are illustrated by a hypothetical example of pumping in a desert basin with a through-flowing river (fig. 33). The basin is 20 mi wide and 40 mi long with a well-connected river entering the basin at the northeast corner, running along the east side of the basin, and exiting at the southeast corner of the basin. Mountain-front recharge of 500 acre-ft/yr is uniformly distributed along the western boundary of the basin. Hydraulic conductivity is 50 ft/d and aquifer thickness is about 500 ft, resulting in a transmissivity of about 25,000 ft²/d. Specific yield is 0.2. The effects of pumping at two possible well locations are shown in figure 34—well 1 is 5 mi from the river, and well 2 is 10 mi from the river. Pumping at either well is at a rate of 600 acre-ft/yr for a period of 50 years, after which pumping ceases. The purpose of this analysis is to better understand the effects of pumping at these locations individually, not simultaneously.

Depletion was calculated by using a groundwater-flow model. The U.S. Geological Survey (USGS) computer program MODFLOW–2005 (Harbaugh, 2005) was used with a single layer with 40 rows and 40 columns of finite-difference cells. Cell dimensions in the east-west direction are 2,640 ft and in the north-south direction are 5,280 ft. A steady-state solution was run prior to simulating pumping. In that solution, the gradient of the river resulted in inflow to the aquifer at a rate of 1,795 acre-ft/yr from the upper part of the river.

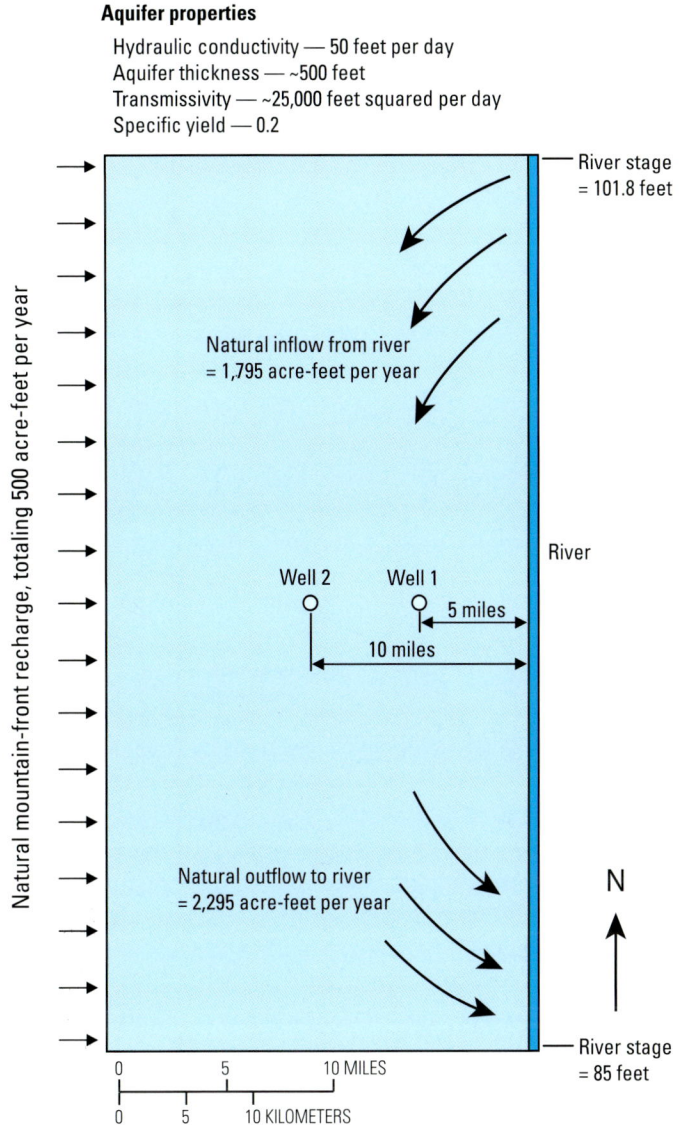

Figure 33. Hypothetical desert-basin aquifer with a through-flowing river along the east side of the basin. In separate analyses, water is pumped at locations of well 1 and well 2 at a rate of 600 acre-feet per year for 50 years.

That inflow, plus 500 acre-ft/yr of mountain-front recharge resulted in 2,295 acre-ft/yr of outflow from the aquifer to the lower part of the river. With this flow pattern, pumping at either location will increase the inflow from the river to the aquifer and decrease the outflow from the aquifer to the river. Pumping cannot, however, increase the specified amount of mountain-front recharge. For cases of pumping at each well location, 50 years of pumping was followed by 100 years of recovery.

Results for pumping at the location of well 1 are shown in the upper three graphs in figure 34. Pumping causes inflow from the river to increase from 1,795 acre-ft/yr to a maximum of 2,009 acre-ft/yr at a time of 50 years. In the 100 years that follow, inflow from the river gradually decreases to

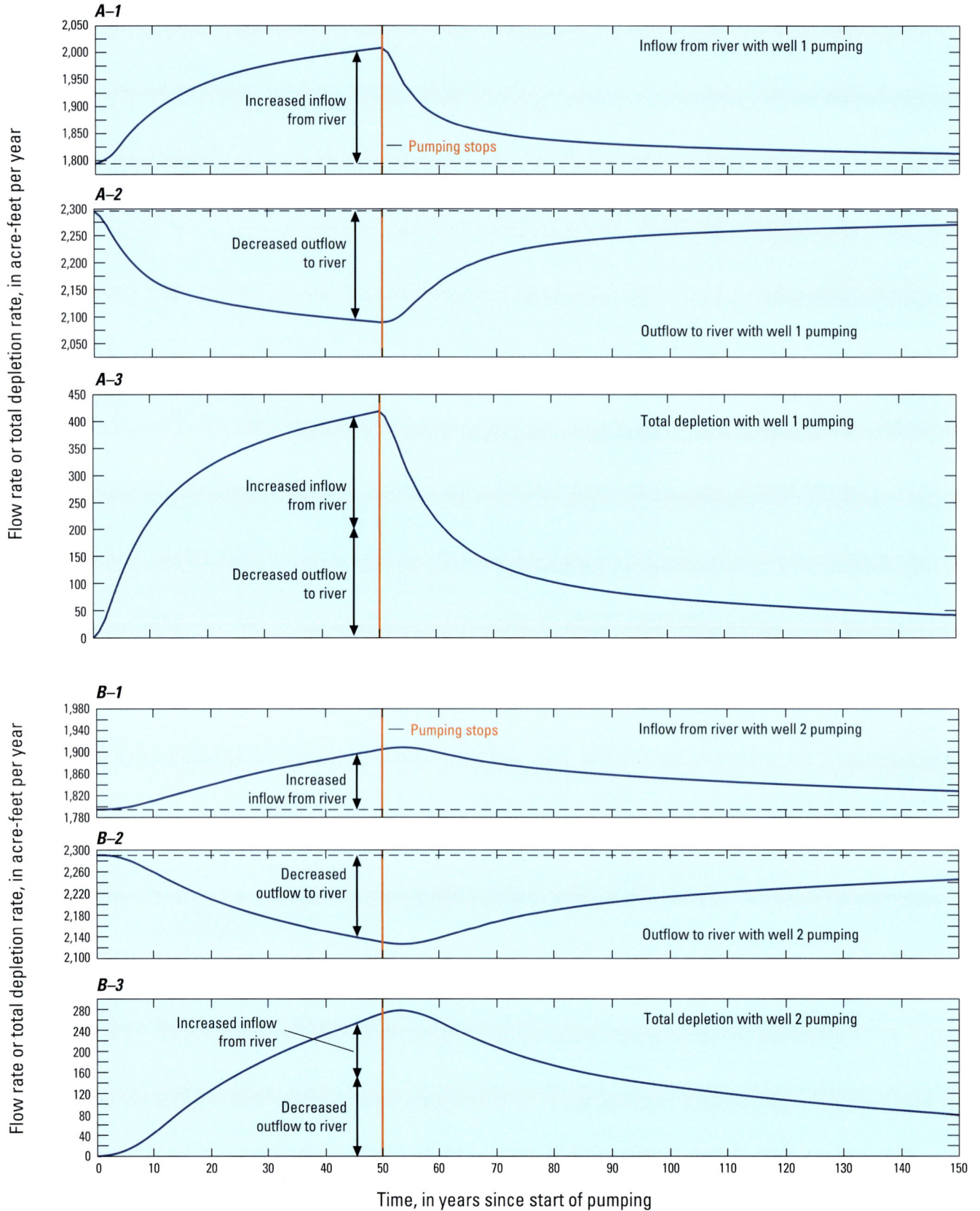

Figure 34.   Inflow from the river, outflow to the river, and total depletion rate with well 1 pumping (upper three graphs) and with well 2 pumping (lower three graphs). Well locations in relation to the river are shown in figure 33.

1,813 acre-ft/yr. Additional time would be required to a full recovery of an inflow value of 1,795 acre-ft/yr. Similarly, pumping causes outflow to the river to decline from 2,295 acre-ft/yr to 2,090 acre-ft/yr at a time of 50 years. In the 100 years that follow with no pumping, outflow increases to 2,270 acre-ft/yr. For any given time, the sum of the increase in inflow from the river and decrease in outflow to the river is the total depletion from pumping. That value begins at zero, reaches a maximum of 419 acre-ft/yr at 50 years, and diminishes to 43 acre-ft/yr at 150 years.

Results for pumping at the location of well 2 are shown in the lower three graphs in figure 34. Pumping causes inflow from the river to increase from 1,795 acre-ft/yr to a maximum of 1,908 acre-ft/yr at a time of 54 years (4 years after pumping stops). In the 96 years that follow, inflow from the river gradually decreases to 1,828 acre-ft/yr. Similarly, pumping causes outflow to the river to decline from 2,295 acre-ft/yr to 2,131 acre-ft/yr at a time of 53 years. In the 97 years that follow with no pumping, outflow increases to 2,250 acre-ft/yr. Total depletion increases from zero to a maximum of 278 acre-ft/yr at 53 years and diminishes to 78 acre-ft/yr at 150 years.

In pumping at either location, 30,000 acre-ft of water is pumped over the 50-year period of pumping (fig. 35). For pumping at the location of well 1, total depletion in the 50-year period of pumping was 15,412 acre-ft, which means that nearly half of the total volume of depletion (30,000 acre-ft) will occur after pumping stops. In contrast, the total volume depleted in 50 years from pumping at the location of well 2 is 7,390 acre-ft, which means that about three-fourths of the total volume of depletion will occur after pumping stops. For pumping at either location, ultimate depletion of 30,000 acre-ft has not occurred in the 150-year period shown (fig. 35), but the trend in the depletion-volume curves is toward that ultimate value.

Most of the six key points listed previously are illustrated by this example. For pumping at the location of well 2, the maximum depletion rate occurred 3 years after pumping stopped. The time interval between the end of pumping and the time of maximum depletion rate will increase with increasing distance between pumping location and connected surface-water features. In the case of the C-aquifer analysis presented previously in the report (figs. 17–19), pumped wells were more than 20 mi from connected surface-water features. Maximum depletion in that analysis was computed to occur about 44 years after pumping stopped (fig. 19; Leake and others, 2005). In addition, the example in this section shows that pumping at either location can both increase inflow from the river and decrease outflow to the river. The sum of these two components is depletion, which represents the total reduction in surface-water flow at any given time.

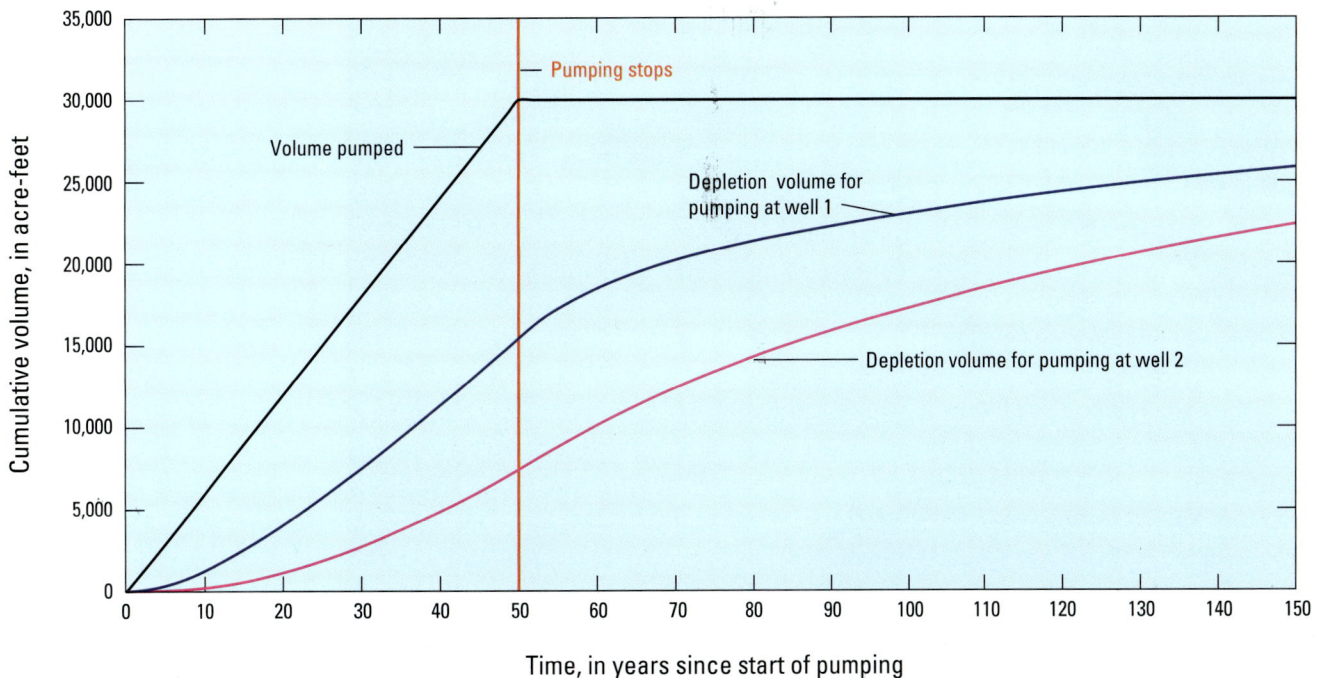

Figure 35.   Cumulative volume pumped and cumulative volume of streamflow depletion for pumping at wells 1 and 2. Well locations in relation to the river are shown in figure 33.

Effects of Confining Layers on Depletion

Various geologic features that act as conduits or barriers to groundwater flow can affect the timing of depletion from groundwater pumping and also can affect which streams are affected by the pumping. Confining layers within or adjacent to aquifers are the most common type of geologic feature that potentially affect timing and locations of depletion. Here the term "confining layers" is used to refer to horizontal or nearly horizontal beds of clay, silt, or other geologic strata that have substantially lower hydraulic conductivity than adjacent aquifer material. In unconsolidated sediments that typically are a part of stream-aquifer systems, aquifer material generally consists of sand and gravel, and confining material generally consists of silt and clay. Confining layers may be laterally discontinuous or they may form laterally extensive barriers that separate adjacent aquifers. Drawdown from a pumped well propagates more rapidly in coarse-grained aquifer material than in confining layers, and in most cases confining layers between pumping locations and streams slow down the progression of depletion in comparison to equivalent aquifer systems without confining layers. It is not reasonable, however, to expect that pumping beneath an extensive confining layer will eliminate depletion. Water does move vertically from one aquifer to another through confining layers, and drawdown from pumping can propagate through confining layers as well. Also, the effective storage coefficient in confined aquifers (beneath confining layers) commonly

Photograph by Stanley Leake, U.S. Geological Survey

Groundwater from aquifers beneath the Colorado Plateau is shown discharging at Fossil Springs in north-central Arizona.

is 2–4 orders of magnitude less than in shallow unconfined aquifers with storage properties dominated by specific yield. Smaller storage coefficients result in faster lateral propagation of drawdown from pumping locations to distant edges of confining layers or locations where drawdown can more easily propagate upward. The argument that pumping beneath a confining layer eliminates the possibility of depletion implies that the pumped aquifer is without any vertical or lateral connection to aquifer material that is connected to surface water. The existence of gradients of water levels in confined aquifers, however, is evidence that the aquifers receive water from and discharge water to vertically adjacent aquifers. Drawdown from pumping also can propagate to these adjacent aquifers. The timing of depletion in systems with extensive confining layers is best understood using numerical models of groundwater flow.

Discontinuous confining layers between pumping locations and connected streams can either slow down or speed up the progression of depletion, depending on the configurations of the confining layers in relation to connected streams and pumping locations. These effects are illustrated using a finite-difference model of the hypothetical basin-fill aquifer shown in figure 36. The aquifer is 30 mi wide, 45 mi long, and 600 ft thick. A river connected to the upper part of the aquifer is present along the center of the basin. Horizontal and vertical hydraulic conductivity, specific yield, and specific storage for coarse sediments and confining clay layers (fig. 36D) are within ranges of values for these types of sediments in real aquifer systems. The larger storage property, specific yield, applies only at the upper boundary of the system where lowering of the water table causes pore spaces to drain. In the aquifer below the water table, a much smaller storage property consisting of the product of specific storage and aquifer thickness accounts for storage changes from compressibility of water and the matrix of solids that makes up the aquifer. Three cases with different configurations of clay layers in the aquifer are shown in figure 36B. In Cases 2 and 3, clay layers are 5 percent of the total aquifer thickness and are near the vertical center of the aquifer.

Horizontal dimensions of finite-difference cells were 1,575 ft in each direction, resulting in 101 columns and 151 rows to simulate the basin width and length, respectively. Twenty layers, each with a thickness of 30 ft, were used to simulate the entire aquifer thickness. Depletion fractions from pumping at four locations in section $A–A'$ at a rate of

1,000 ft^3/d for 25 years were computed using the superposition modeling approach with MODFLOW–2005 (Harbaugh, 2005).

Comparison of depletion curves for the three cases and four pumping locations (fig. 37) yields some insights into the range of effects of clay layers on depletion. The first result to note is that even with no clay layer present, depletion from pumping at depth in some locations progresses faster than depletion from pumping near the top of the aquifer. For example, with no clay layer, depletion progresses slightly faster from pumping at depth (fig. 37B) than from pumping nearer to the water table (fig. 37A). This difference occurs because vertical hydraulic conductivity is much lower than the horizontal hydraulic conductivity. Drawdown from pumping at depth can propagate more easily laterally toward the river location than to the overlying water table where the specific yield value can result in large storage-change values that slow the propagation of the cone of depression.

The existence of a clay layer under the river (Case 2) greatly slows depletion for the deep pumping location nearer to the river (fig. 37D). The clay layer restricts direct propagation of drawdown upwards to the river. Drawdown must propagate laterally around the edge of the clay layer and then back to the river. This case is similar to the situation in the Upper San Pedro Basin in Arizona, where a silt and clay layer underlies the stream at most locations (fig. 13).

The existence of clay layers along the margins of the valley (Case 3) substantially speeds up the depletion for the pumping location beneath that layer (fig. 37B). The clay layer speeds up depletion from underlying pumping because it creates a confined aquifer zone that restricts propagation of drawdown to the water table and, with a small storage coefficient, allows relatively rapid propagation of drawdown to the edge of the clay layer.

In summary, confining layers and other geologic features are complexities that can affect the timing of depletion from groundwater pumping. If features have a lower hydraulic conductivity than that of aquifer material, the feature can slow down the progress of depletion through time. In some cases, such as is shown in figure 37B, the feature may speed up the progress of depletion. For systems with multiple aquifers separated by confining layers, or for aquifers with discontinuous confining layers and other heterogeneities, numerical flow modeling approaches are needed to better understand the timing of depletion.

A. Model dimensions and location of section *A–A′*

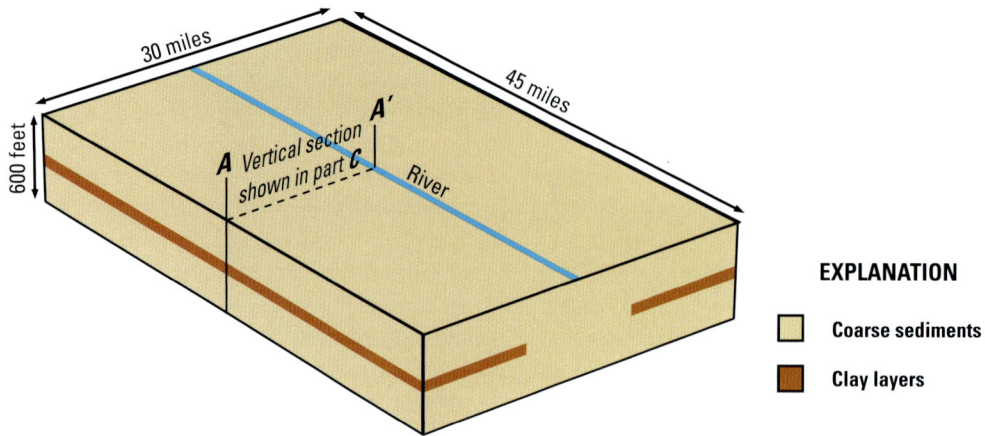

B. Configurations of clay layers for Cases 1, 2, and 3

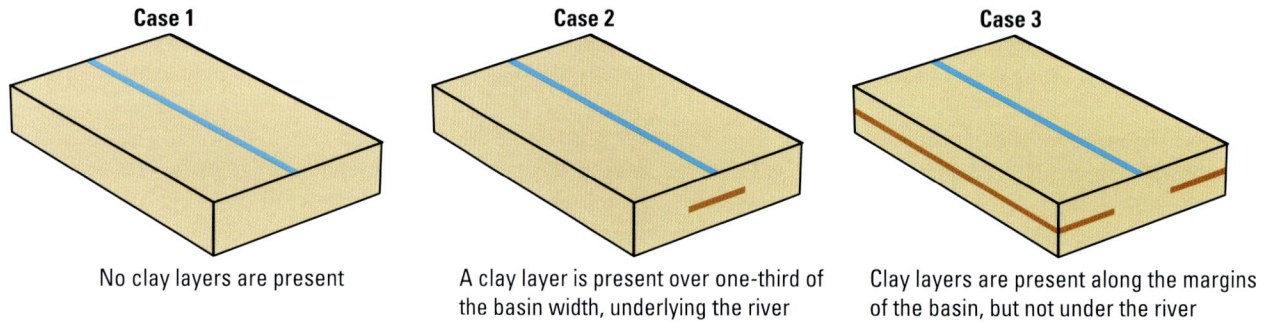

C. Geometry of section *A–A′*

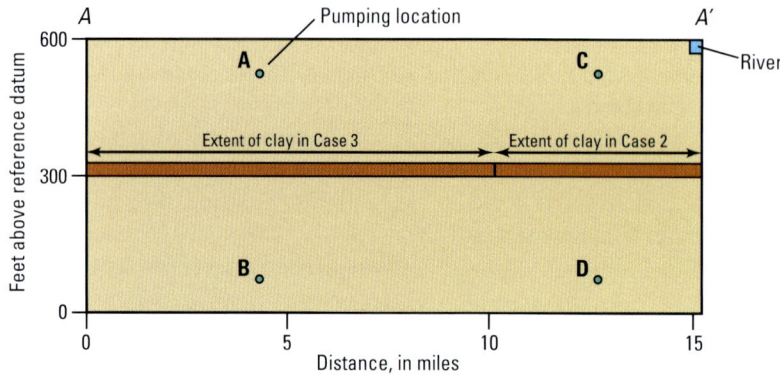

D. Aquifer properties

	Horizontal hydraulic conductivity, in feet per day	Vertical hydraulic conductivity, in feet per day	Specific storage, in per foot	Specific yield, dimensionless
Coarse sediments	3×10^1	3×10^{-1}	6×10^{-7}	2×10^{-1}
Clay layers	3×10^{-3}	3×10^{-5}	6×10^{-7}	2×10^{-1}

Figure 36. *A*, Hypothetical basin-fill aquifer used to illustrate possible effects of discontinuous clay layers on timing of depletion in the river as a function of vertical and horizontal locations of pumping. *B*, Configurations of clay layers are shown for three cases. *C*, Depletion in vertical section *A–A′* is shown in figure 37 for pumping locations A, B, C, and D. *D*, Aquifer properties are within the range of values typical of basin-fill aquifers, with a horizontal-to-vertical hydraulic conductivity ratio of 100:1. Clay layers in Cases 2 and 3 increase restrictions to vertical flow in parts of the aquifer.

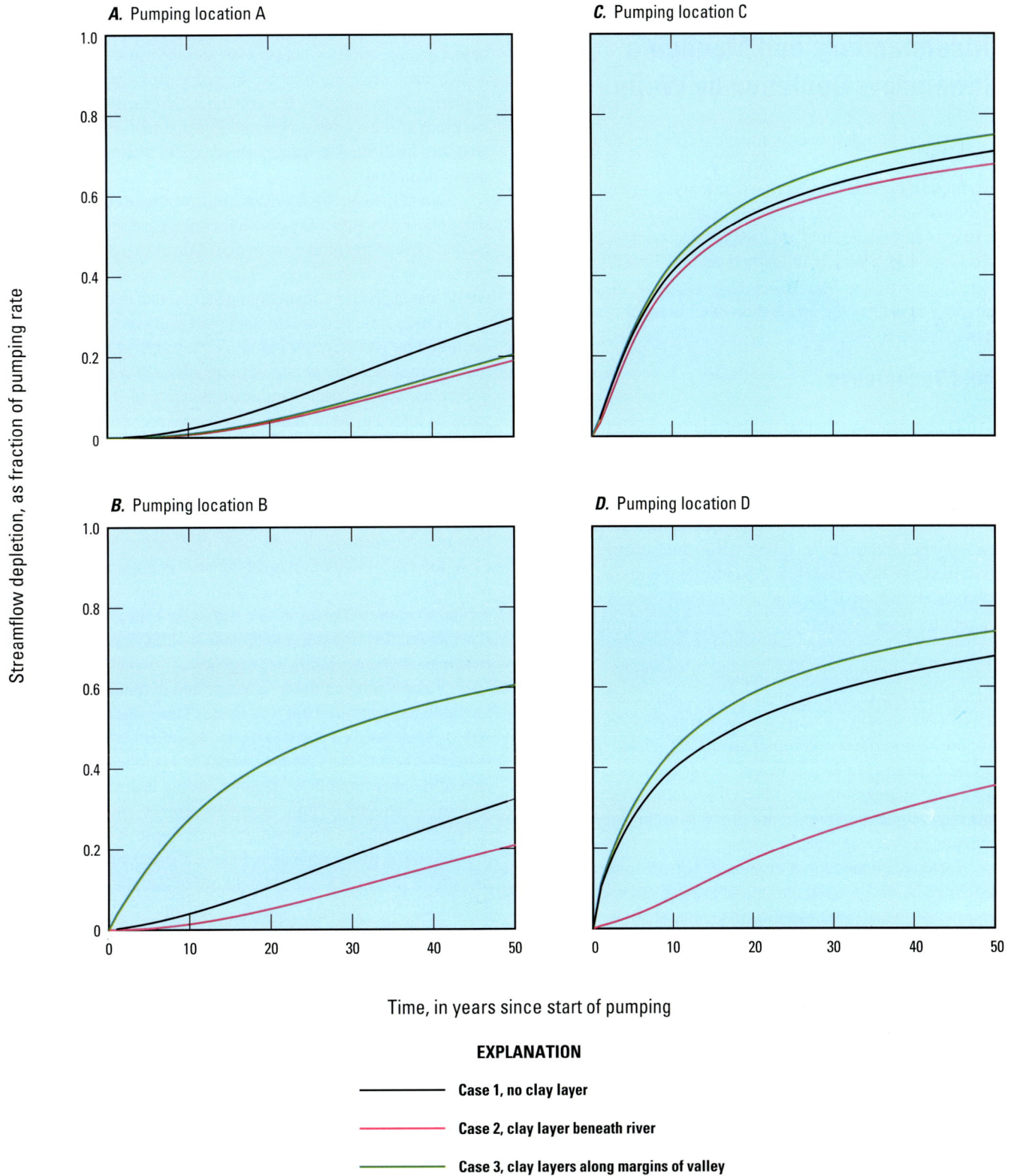

Figure 37. Computed depletion at pumping locations A, B, C, and D in vertical section *A–A'* shown in figure 36*C*. For *A*, shallow distant pumping location A, either configuration of clay layers slows depletion in comparison to case 1. For *B*, deep distant pumping location B, pumping below the clay layer at the valley margins (Case 3) produces substantially more rapid depletion than in the case with no clay layers. For *C*, shallow close pumping location C, configurations of clay layers change depletion from the case of no clay layer by a minor amount. For *D*, deep close pumping location D, the clay layer beneath the river (Case 2) substantially slows the process of depletion.

Approaches for Monitoring, Understanding, and Managing Streamflow Depletion by Wells

This section describes approaches for determining the effects of pumping on streamflow. These approaches fall into two broad categories: collection and analysis of field data and analytical and numerical modeling. Additionally, this section describes approaches that are used for managing streamflow depletion, which build on both an understanding of the underlying processes that affect the response of streamflow to pumping as well as the application of techniques for modeling these processes.

Field Techniques

Quantification of streamflow depletion using field-measurement techniques first requires that changes in flow between the stream and aquifer can be measured or estimated. The measurement technique must have the capability to resolve changes in streamflow that occur over a stream reach that may be affected by a well or wells. Such changes can most likely be detected when groundwater pumping is a substantial fraction of the available streamflow and when enough time has elapsed since pumping began for depletion to occur. A second requirement is that there must be a way to separate pumping-induced changes in streamflow from changes in flow caused by other stresses such as climate-driven variations in recharge and stream stage. Changes in groundwater/surface-water interactions not related to pumping can be as large as or larger than pumping-induced changes. Separating pumping-driven effects from other effects in field data may require comprehensive analysis of the coupled groundwater/surface-water system.

Field techniques for determination of streamflow depletion can be grouped into the following general approaches, which are described in greater detail by Stonestrom and Constantz (2003), Rosenberry and LaBaugh (2008), and Constantz (2008):

1. Direct measurement of streamflow,

2. Point measurements of flow across the streambed, or

3. Measurement of other types of data that indicate the direction or quantity of flow between a stream and adjoining aquifer.

The second approach includes seepage meters placed at specific points in the stream channel (Rosenberry and LaBaugh, 2008). Many of the additional data types in the third approach also focus on specific point measurements in a stream channel but also include methods that monitor larger areas of a stream reach. These approaches employ water levels measured at observation wells or streambed piezometers, measurements of temperature in the stream and streambed,

analysis of geochemical constituents or tracers, and geophysical studies of the stream-aquifer system. Field methods that detect changes in flow between an aquifer and a stream over a long reach are more likely to be successful in detecting depletion from pumping than methods that focus on specific locations along a stream channel. For that reason, this discussion will be limited to direct measurement of streamflow to detect depletion.

Direct measurements of streamflow are used to determine streamflow changes that occur either at a particular stream location over time or at a particular time at two or more locations along a stream. Repeated streamflow measurements at a single site such as a streamgaging station can detect changes in flow over time that are driven by all processes, including depletion by groundwater pumping. Streamflow measurements made simultaneously at two or more sites along a stream are known as "seepage runs" and are indicative of streamflow gains or losses in the reaches between measurement locations. Detection of depletion using seepage runs requires that two or more measurements be made during a period in which substantial streamflow depletion may occur.

The identification of streamflow depletion from streamflow measurements is complicated by a number of factors. First, the rate of depletion must be large enough to be detected by the streamgaging instrumentation, and significantly greater than the accuracy of the streamflow measurement. Each streamflow measurement made by USGS personnel, for example, is given a rating of "excellent," "good," "fair," or "poor," depending on the hydrologic and hydraulic conditions in which the measurement was made (Turnipseed and Sauer, 2010). As defined by the USGS, an "excellent" rating is one in which the accuracy of the measurement is judged to be ±2 percent of the measured flow, a "good" rating is one with an accuracy of ±5 percent, a "fair" rating ±8 percent, and a "poor" rating greater than ±8 percent. As an example of the effects of streamflow-measurement accuracy, a streamflow-depletion rate of 1.6 ft³/s (1.0 Mgal/d) could not be accurately detected using the USGS rating system for a stream with a measured flow of 100 ft³/s, even if the streamflow measurement had been rated as "excellent." This is because the depletion rate of 1.6 ft³/s is less than the 2.0 ft³/s accuracy of the measurement.

A related complicating factor is the effect that the aquifer has on delaying the time of arrival and on damping the range of streamflow-depletion rates caused by pumping at a well. Many of the examples provided in previous sections of this report, such as that for the Upper San Pedro Basin of Arizona (fig. 13), have demonstrated that it may take several years, if not decades, for a pumping stress to be manifested in a stream. The propensity of the aquifer to delay and damp a particular pumping stress can make it extremely challenging, if not impossible, to monitor streamflow depletion in some field settings or to differentiate streamflow depletion caused by pumping at a particular well or well field from depletion caused by other short-term or long-term stresses to the aquifer (Zlotnik, 2004; Bredehoeft, 2011b).

Left, streamgage on the Snake River, near Moran, Wyoming. (Photograph from U.S. Geological Survey files)

Streamflow measurement on Fish Creek, Teton County, Wyoming. (Photograph by Jerrod D. Wheeler, U.S. Geological Survey)

Above, seepage meters and in-stream piezometers deployed in the Shingobee River, Minnesota, to understand directions and rates of water movement between the stream and the underlying groundwater system. (Photograph by Donald O. Rosenberry, U.S. Geological Survey)

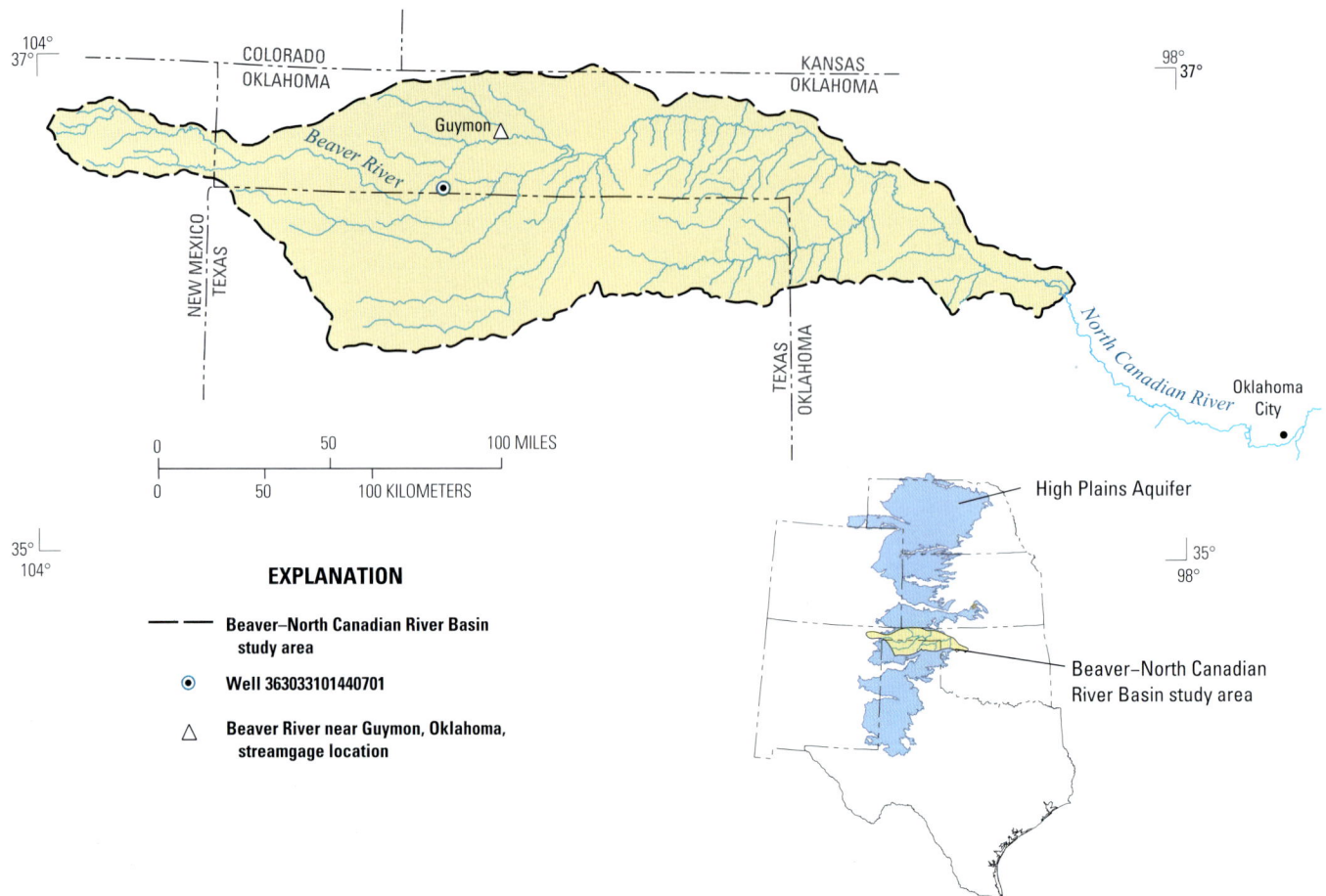

Figure 38.   Location of the Beaver–North Canadian River Basin, western Oklahoma (modified from Wahl and Tortorelli, 1997).

In light of these challenges, techniques for monitoring streamflow depletion have been limited to two general approaches. In the first, short-term field tests lasting several hours to several months are done to determine local-scale effects of pumping from a specific well or well field on streams that are in relatively close proximity to the location of withdrawal. In the second, evaluations are made of hydrologic and climatic data collected over a period of many years to determine whether changing streamflow conditions can be correlated to long-term, basinwide development of groundwater resources (Wahl and Tortorelli, 1997; Sophocleous, 2000; Fleckenstein and others, 2004; and Prudic and others, 2006). Analyses of this type typically use statistical techniques to identify and explain long-term trends in streamflow conditions.

Short-term tests to determine local-scale effects of pumping are done for two primary purposes. The first is to determine the effects of an existing production well (or well field) on specific stream reaches or, conversely, to determine the quantity (and often the quality) of surface water captured by a well. Examples of these types of studies are provided by Myers and others (1996) for a site in Kansas and Dudley and Stewart (2006) for a site in Maine, and by several studies cited previously related to bank filtration. The second purpose is to

improve the general scientific understanding of the geologic and hydrogeologic controls on streamflow depletion or to test the predictive ability of analytical and numerical models to determine streamflow depletion under actual field conditions. Field studies such as these typically make use of multiple data types to provide a comprehensive picture of how stream-aquifer systems respond to pumping. Experimental studies of this type include those described by Sophocleous and others (1988), Christensen (2000), Hunt and others (2001), Nyholm and others (2002, 2003), Hunt (2003b), Kollet and Zlotnik (2003), Fox (2004), and Lough and Hunt (2006).

An example of a study in which long-term changes in streamflow were correlated with groundwater development is provided for the Beaver–North Canadian River Basin of western Oklahoma (fig. 38). The basin is underlain by the High Plains aquifer, which is one of the most productive aquifer systems of the United States. Groundwater levels have declined in many parts of the High Plains aquifer in response to large-scale development of groundwater for irrigation that began in the 1940s (McGuire and others, 2003). These declines are illustrated by water levels measured in an observation well in western Oklahoma since the 1950s (fig. 39A). The study was undertaken in response to concerns about streamflow reductions that seemed to be occurring in the

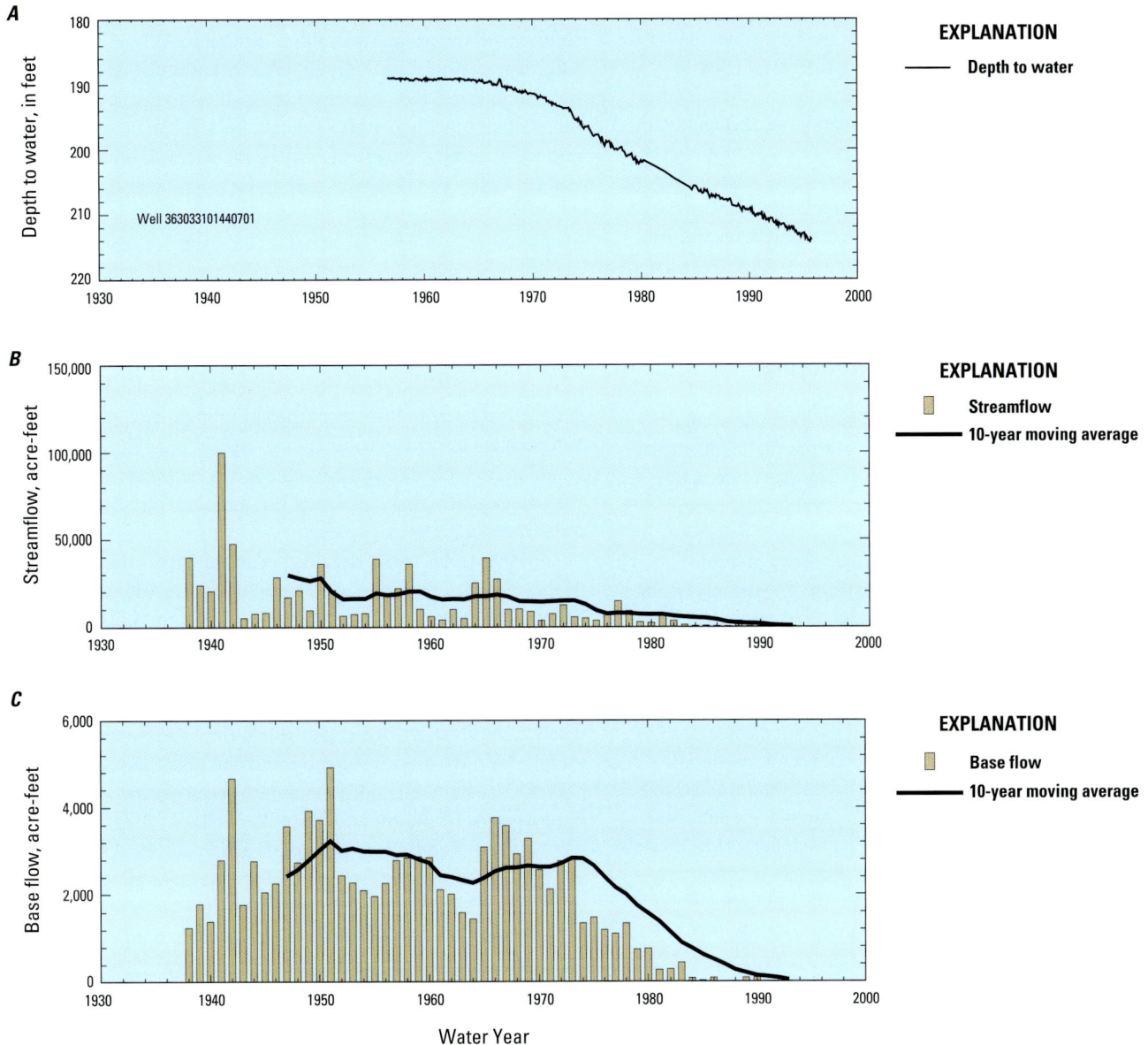

Figure 39. Long-term hydrologic data for the Beaver–North Canadian River Basin, western Oklahoma. *A*, Groundwater levels in an observation well in Texas County (1956–95). *B*, Total annual volume of streamflow and, *C*, base flow for the Beaver River near Guymon, Oklahoma (1938–93; modified from Wahl and Tortorelli, 1997). Location of observation well and streamgaging station shown in figure 38.

North Canadian River, which at the time of the study was the source of about half of the public-water supply for Oklahoma City (Wahl and Tortorelli, 1997).

Several sources of hydrologic and climatic data were analyzed as part of the study, including the measured volume of annual streamflow and estimated volume of annual base flow at several USGS streamgaging stations in the basin. The data were divided into an "early" period (ending in 1971), representing conditions before groundwater levels had declined substantially, and a "recent" period (1978–94), reflecting the condition of declining groundwater levels (Wahl and Tortorelli, 1997). Statistical tests of the data showed

that the total volume of annual streamflow measured at most of the streamgaging stations in the basin had decreased from the early to recent periods, even though precipitation records for the area showed no corresponding changes. Groundwater discharge to streams in the basin had also undergone significant changes, with substantial reductions documented at some of the streamgaging stations. These trends are illustrated by streamflow data and base-flow estimates for the Beaver River at Guymon, Oklahoma (fig. 39B and C). Overall, the observed reductions in streamflow throughout the basin correlated well with long-term declines in groundwater levels that occurred in response to increased pumping for irrigation, although other factors such as changes in farming and conservation practices in the basin also may have had an effect on the changes in streamflow (Wahl and Tortorelli, 1997).

Statistical studies such as these can be used in general to evaluate the large-scale effects of basinwide pumping on streamflow reductions. They cannot, however, account for the specific effects of pumping at individual wells, nor can they help with understanding how specific management actions might affect future depletion. Such analyses require the use of analytical or numerical models.

Analytical and Numerical Modeling

Analytical and numerical modeling methods are the most widely applied approaches for estimating the effects of groundwater pumping on streamflow. The two approaches use different mathematical techniques to solve the partial differential equation of groundwater flow (or change in groundwater flow). The groundwater-flow equation mathematically describes the distribution of hydraulic heads (or drawdowns) throughout a groundwater system over time. Analytical models are limited to the analysis of idealized conditions in which many of the complexities of the real groundwater system are either ignored or approximated by use of simplifying assumptions. These simplifications typically include representation of the three-dimensional flow system by a one- or two-dimensional system, idealized boundary conditions such as perfectly straight streams, and homogeneous aquifer materials. In contrast, numerical models are capable of simulating fully three-dimensional flow in groundwater systems that are horizontally and vertically heterogeneous and have complex boundary conditions.

Although both modeling approaches have been widely used, numerical models provide the most robust approach for determining the rates, locations, and timing of stream-flow depletion by wells. Nevertheless, because analytical models have received substantial application and continue to be the subject of much research, a brief review of the history and scope of analytical solutions for analysis of streamflow depletion is provided. Different approaches for numerical-modeling analyses of streamflow depletion also are described and provide background for the discussions on streamflow-depletion response functions, capture maps, and management approaches.

Analytical Models of Streamflow Depletion by Wells

Several analytical solutions to the groundwater-flow equation have been developed to determine time-varying rates of streamflow depletion caused by pumping. Analytical solutions are based on highly simplified representations of

field conditions that are necessary to develop mathematical solutions to the groundwater-flow equation. Although these solutions are highly simplified representations of real-world field conditions, they can provide insight into the several factors that affect streamflow depletion and can be used as an initial estimate of the effects of a particular well on a nearby stream. Partly because they require less site-specific data to implement than do numerical models, analytical models have been used by a number of States as the basis for making water-management regulatory decisions (Sophocleous and others, 1995; Miller and others, 2007; Reeves and others, 2009).

Figure 40*A* illustrates a hypothetical stream-aquifer system that is representative of many river-valley aquifers of the United States. A single well pumps from the aquifer and captures streamflow from the adjoining major stream and perhaps also from tributaries to the stream. The aquifer is underlain by sediments having a lower permeability than the aquifer (such as glacial till) and then by relatively impermeable bedrock.

Many, if not most, streams penetrate only a small fraction of the saturated thickness of the adjoining aquifer, such as illustrated for the stream in figure 40*A*. This condition is referred to as a partially penetrating stream, and both the stream and the well in figure 40*A* partially penetrate the aquifer. Partially penetrating streams and pumped wells can create complicated three-dimensional flow patterns in the vicinity of the wells and streams and can result in water being captured by the wells from parts of the aquifer that are on the opposite side of the streams from the wells.

A simplified representation of the hypothetical river-valley aquifer is shown in figure 40*B*. The conceptualization of the stream-aquifer-well system forms the basis for the development of the simplest and most widely applied analytical solution of streamflow depletion, which was developed independently and in somewhat different forms by Theis (1941) and Glover and Balmer (1954) and later implemented in a set of tables and graphs by Jenkins (1968a). The solution is based on several assumptions, including representation of the partially penetrating stream by one that fully penetrates the aquifer. Other assumptions are that (1) the aquifer is confined, homogeneous, underlain by an impermeable boundary, and extends to infinity in all directions away from the stream; (2) the aquifer is bounded by a single stream that is straight and in perfect hydraulic connection with the aquifer (that is, there are no resistive streambed sediments at the stream-aquifer interface); and (3) a single well pumps from the full saturated thickness of the aquifer. The solution is frequently applied to unconfined aquifers (as in figure 40) when it can be assumed that drawdowns caused by pumping at a well are small compared to the initial saturated thickness of the aquifer. The solution is sometimes referred to as the "Glover solution" or "Jenkins' approach" and, because it has been so widely applied, is discussed in more detail in Box C.

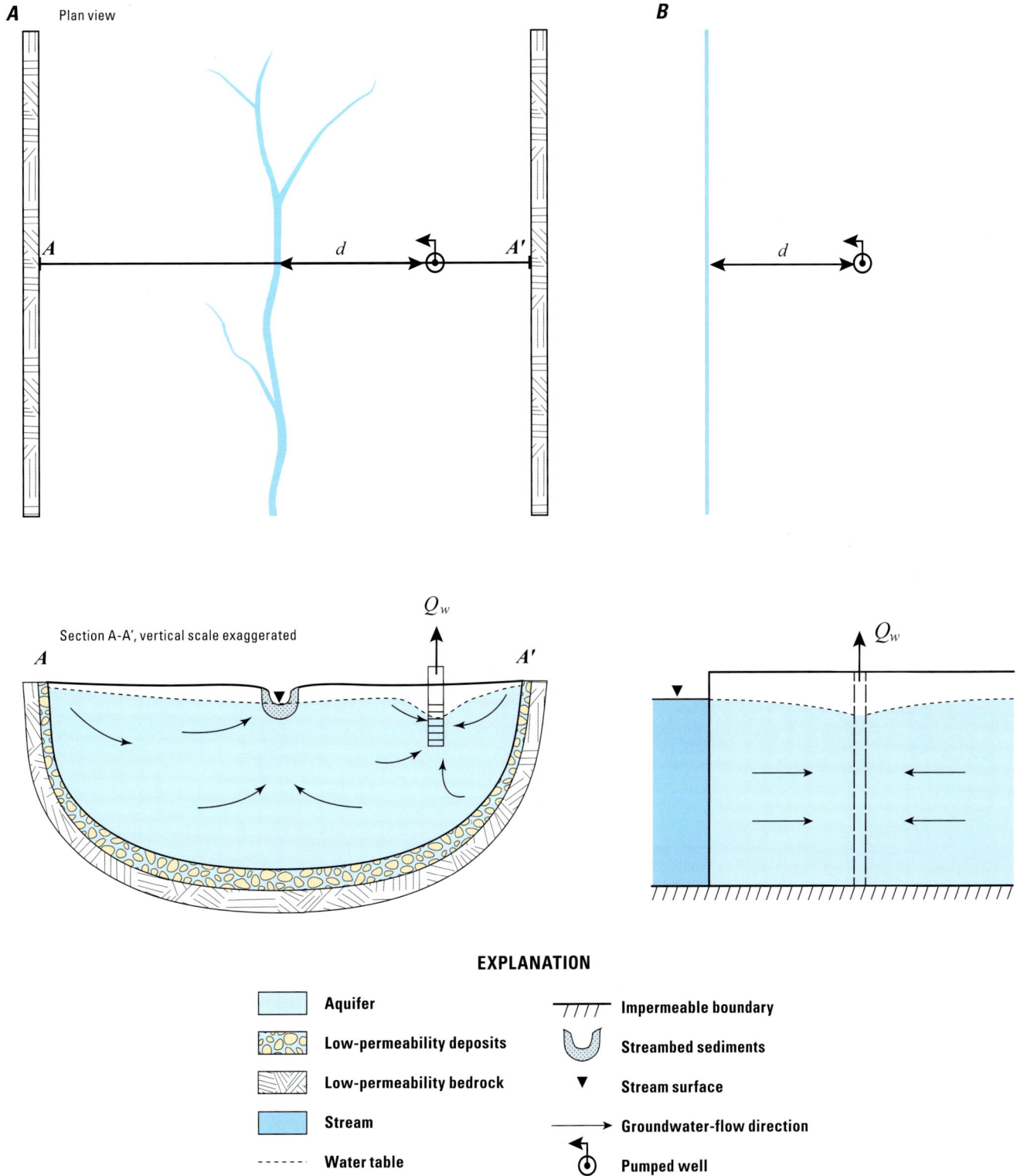

Figure 40. *A*, Hypothetical river-valley aquifer with a single pumping well. *B*, Simplified conceptualization of the same river-valley aquifer for the Glover analytical solution. [*d* is distance from well to nearest stream and Q_w is pumping rate at well]

Box C: Glover's Analytical Solution and Jenkins' Stream Depletion Factor (SDF)

The most widely applied analytical solution for determining the effects of pumping on streamflow is one that was developed by Glover and Balmer (1954) that has become known as the Glover solution. The solution is based on a highly simplified stream-aquifer-well system illustrated in figure 40B. The full set of assumptions on which the solution is based can be summarized as follows (Jenkins, 1968a):

1. The aquifer is homogeneous, isotropic, and extends to infinity away from the stream.

2. The aquifer is confined, and the transmissivity and saturated thickness of the aquifer do not change with time. The solution is also applied to water-table aquifers when it can be assumed that drawdown caused by pumping is small compared to the initial saturated thickness of the aquifer.

3. Water is released instantaneously from storage (and there are no delayed-drainage effects characteristic of water-table aquifers).

4. The stream that forms a boundary with the aquifer is straight, fully penetrates the thickness of the aquifer, is infinitely long, remains flowing at all times, and is in perfect hydraulic connection with the aquifer (that is, no streambed and streambed sediments impede flow between the stream and aquifer).

5. The temperature of the stream and aquifer are the same and do not change with time. This assumption is necessary because variations in temperature affect the hydraulic conductivity of streambed and aquifer sediments.

6. The well pumps from the full saturated thickness of the aquifer at a constant rate.

The assumption that the aquifer is confined (or that the drawdown in a water-table aquifer is small compared to the initial saturated thickness of the aquifer) in conjunction with the two assumptions that the stream and well penetrate the full saturated thickness of the aquifer imply that groundwater flow in the aquifer is horizontal.

The Glover solution provides an expression for the total rate of streamflow depletion as a function of time (defined mathematically as Q_s), and is equal to the product of the pumping rate of the well, Q_w, and a mathematical function referred to as the complementary error function, $erfc(z)$:

$$Q_s = Q_w erfc(z) . \qquad (C1)$$

Variable z in this equation is equal to $\sqrt{(d^2 S)/(4Tt)}$, in which d is the shortest distance of the well to the stream, S is the storage coefficient of the aquifer (or specific yield, for water-table aquifers), T is the transmissivity of the aquifer, and t is the time. Note that the ratio S/T is the inverse of the hydraulic diffusivity of the aquifer ($D = T/S$). The solution is illustrated in figure C–1A for two wells pumping from an aquifer having a hydraulic diffusivity of 10,000 feet squared per day.

As noted previously in the report, Jenkins (1968a and 1968b) defined the quantity d^2/D, which is equivalent to $(d^2 S)/(T)$ in equation C1, as the "stream depletion factor," or "SDF." Jenkins' SDF has the units of time, such as seconds or days, depending on the units of time used to express T. Although Jenkins named the constant the stream depletion factor, it might alternatively be called a "streamflow-depletion response-time factor" because of its similarity to other types of hydraulic response-time constants that have been defined for groundwater systems (Domenico and Schwartz, 1990; Alley and others, 2002; Sophocleous, 2012).

For the two wells illustrated in figure C–1C, the SDF of well A is 6.25 days and that of well B is 25 days. Note that as either the distance of the well from the stream increases or the hydraulic diffusivity of the aquifer decreases, rates of streamflow depletion increase more slowly. Jenkins also noted that for the stream-aquifer conditions modeled by equation C1, the stream depletion factor is equal to the time at which streamflow depletion is equal to 28 percent of the volume pumped for a given location. This can be seen graphically in figure C–1B, in which the total volume of streamflow depletion for pumping at wells A and B is 28 percent (that is, a fraction of 0.28) of the volume pumped at 6.25 days and 25 days, respectively.

An important aspect of the *SDF* is that it can be calculated for every location in an aquifer. Wells pumping at the same rate and with the same pumping schedule at any location having a particular *SDF* value will have an equal effect on streamflow depletion, assuming that the conditions for which equation C1 were derived are met. Figure C–1*C* illustrates a map of stream depletion factors for the simplified stream-aquifer system that meets the assumptions underlying equation C1. As an example use of the map, any well pumped at a constant rate along the *SDF* contour equal to 25.0 days will result in a streamflow-depletion rate of about 57 percent of the pumping rate of the well after 40 days of pumping (from figure C–1*A*, well B curve) and a total volume of streamflow depletion equal to about 37 percent of the total volume of water pumped to that time (figure C–1*B*, well B curve).

In many field settings, the conditions required for application of the Glover solution and Jenkins' *SDF* mapping approach are not fully met, such as for aquifers that are bounded laterally by low-permeability rocks or sediments. In such cases, methods have been developed to determine modified *SDF* values that account for the specific conditions of the particular field setting (for example, Jenkins 1968b and 1968c; Hurr, Schneider, and others, 1972; Burns, 1983; and Miller and others, 2007). More recently, alternative approaches to the *SDF* methodology have been developed to map aquifer locations having equal effect on streamflow depletion, such as response-function and capture maps.

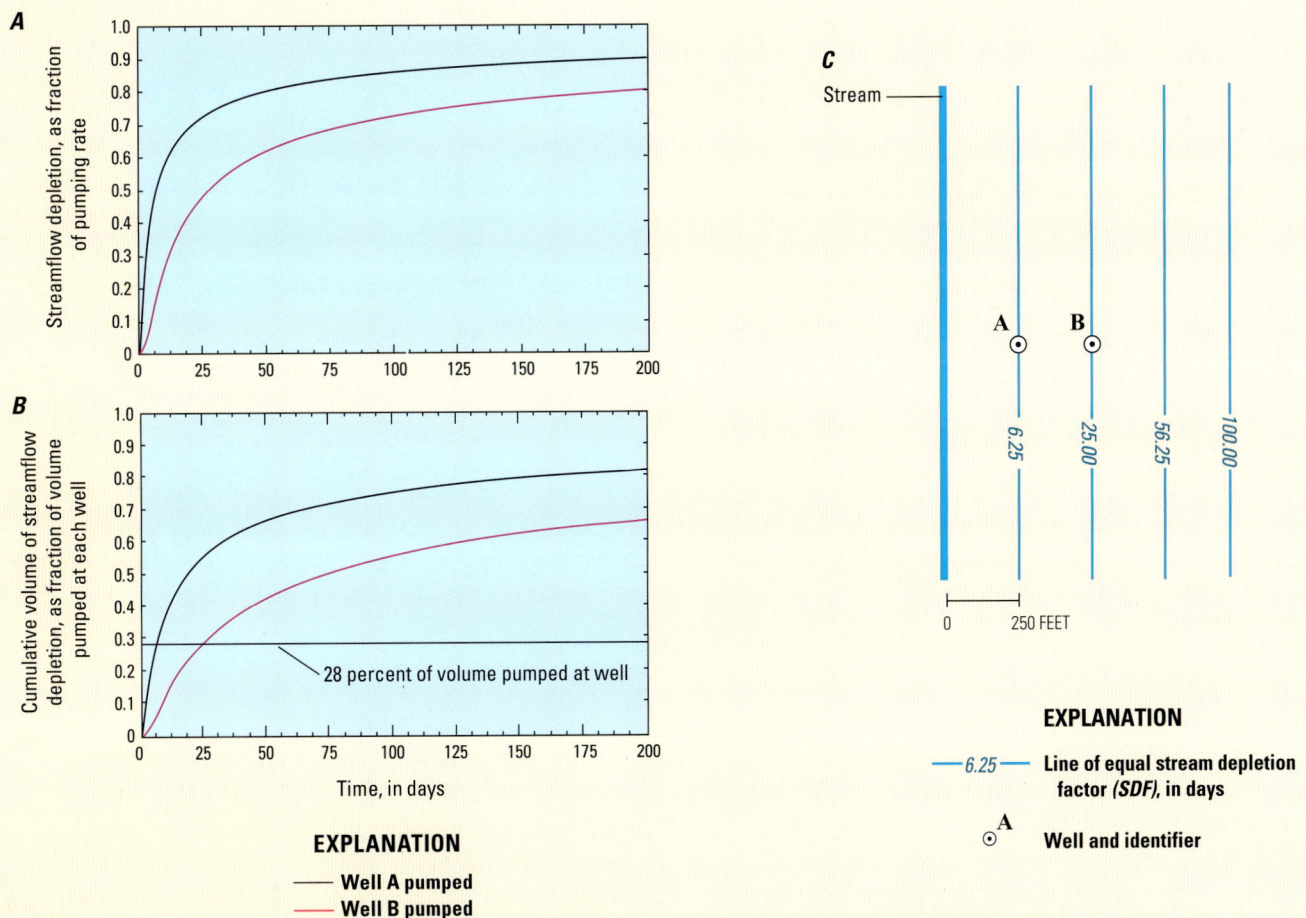

Figure C–1. *A,* Rate and, *B,* cumulative volume of streamflow depletion caused by pumping at two wells located 250 feet (well A) and 500 feet (well B) from a stream. Rates of streamflow depletion were calculated by use of the Glover equation (C1), as implemented in the computer program described in Reeves (2008); cumulative volumes were calculated by adding the daily rates of streamflow depletion. Each well is pumped independently of the other at a rate of 1 million gallons per day from an aquifer having a hydraulic diffusivity of 10,000 feet squared per day. *C,* Contours of stream depletion factor for the aquifer.

A

B

C

EXPLANATION

	Aquifer
	Confining layer
////	Impermeable boundary
	Streambed sediments
---▼---	Water table or stream surface
⟶	Groundwater-flow direction

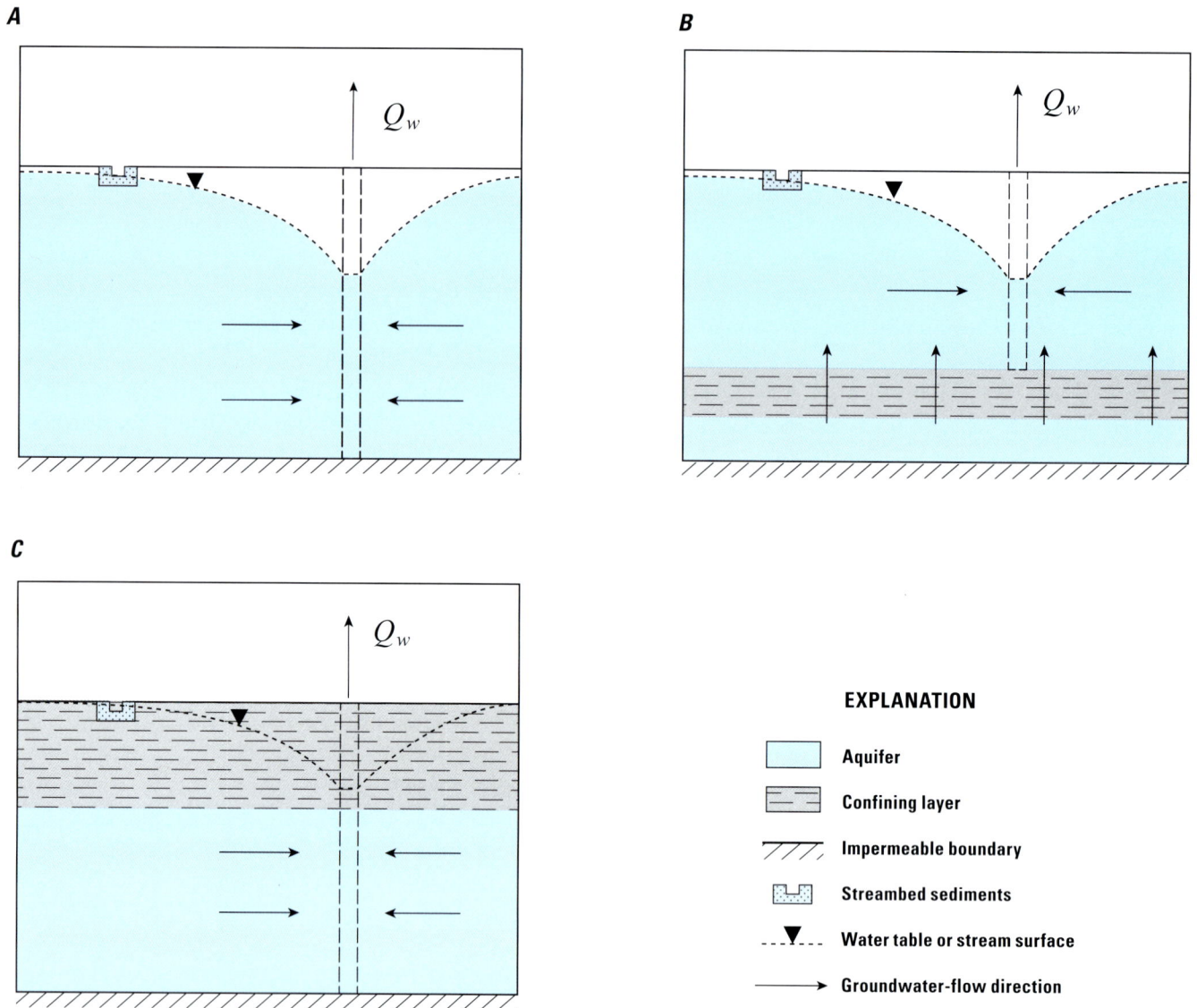

Figure 41.   Alternate conceptualizations of stream-aquifer systems for which analytical solutions have been developed. *A,* Single-layer aquifer with a partially penetrating stream. *B,* Leaky-aquifer system with flow through a low-permeability confining layer from an underlying aquifer. *C,* Leaky-aquifer system with flow from an overlying confining layer (modified from Reeves and others, 2009). [Q_w is pumping rate at well]

Since the initial work of Theis and of Glover and Balmer, many additional solutions have been derived to represent more realistic field conditions. Glover (1974) presented a solution to compute streamflow depletion from a well pumping between a stream and a lateral impermeable boundary that is parallel to the stream. Such an approach would be needed to represent the well pumping between the stream and the impermeable edge of the valley shown in figure 40*A*. Other solutions have focused on incorporating the effects of field conditions such as are shown in figure 41.

Several authors have demonstrated that the assumptions that a stream is in perfect hydraulic connection with the aquifer and extends over the full thickness of the aquifer can lead to significant errors in the determination of the timing and rates of streamflow depletion (Spalding and Khaleel, 1991; Sophocleous and others, 1995; Conrad and Beljin, 1996). Hantush (1965) was the first to develop a solution that accounted for resistance to flow at the stream-aquifer boundary due to streambed materials having a lower hydraulic conductivity than the adjacent aquifer, although his solution was based on a conceptualization of a fully penetrating stream. Hunt (1999), Butler and others (2001), Fox and others (2002), and Singh (2003) later extended the work of Hantush to allow both streambed resistance and partial penetration of the stream into the aquifer (fig. 41*A*). Simulating the stream as partially penetrating the aquifer allows for the propagation of drawdown under the stream and resulting groundwater-storage changes on the side of the stream opposite to the well. An important assumption common to all of these approaches is that the groundwater level in the aquifer at the stream remains above the streambed, such that the stream does not become disconnected from the underlying aquifer.

Additional solutions have been developed to address flow conditions along the lower and upper boundaries of the aquifer. Zlotnik (2004), Butler and others (2007), and Zlotnik and Tartakovsky (2008) developed solutions for leaky-aquifer systems in which the pumped aquifer is underlain by a low-permeability confining layer that restricts flow between the pumped aquifer and an underlying high-permeability aquifer (fig. 41B). In a separate set of papers, Hunt (2003a, 2008) developed analytical solutions for the condition in which the affected stream is located within an overlying confining layer that provides a source of leakage to the underlying pumped aquifer during the early stages of withdrawal (fig. 41C). At equilibrium, however, streamflow depletion is the only source of water to the well.

Although much work has been done to extend the applicability of analytical solutions to conditions that are typically found in the field, these solutions remain unable to address many of the complicating factors that affect streamflow depletion by wells, such as aquifer heterogeneity (Sophocleous and others, 1995; Kollet and Zlotnik, 2003) and the presence of meandering streams that have multiple tributaries. Moreover, even solutions that have been developed to represent aquifers having a finite width (that is, aquifers bounded laterally by low-permeability materials such as shown in figure 40A) are difficult to apply because of irregular geometry of lateral boundaries. It is the authors' experience that these three factors—aquifer heterogeneity, multiple streams and (or) complex stream geometry, and finite-width aquifers with complex geometry—can have substantial effects on streamflow depletion that limit the use of analytical solutions for many practical applications, particularly basinwide analyses in which multiple wells pump simultaneously. For these conditions, numerical-modeling methods are needed.

Winding channel of the Washita River between Anadarko and Chickasha, Oklahoma.

Numerical Models of Streamflow Depletion by Wells

Difficulties in applying analytical approaches to streamflow-depletion problems in real-world settings are apparent in the diagram of a portion of a stream-aquifer system shown in figure 42. Analytical solutions assume a single straight stream, yet the system shown includes a stream and two tributaries, each with irregular geometry. Similarly, analytical solutions would not be able to account for effects of the irregular edges of the aquifer. When faced with these and other complexities, a numerical-modeling approach is needed for analysis of streamflow depletion. Numerical groundwater models are the most powerful tools for understanding streamflow depletion from groundwater pumping. Some of the more common complexities of real systems that require a numerical-modeling approach include

- Irregular geometry of lateral and vertical boundaries of aquifers.

- Irregular geometry of streams, rivers, and other surface-water features.

- Non-uniform (heterogeneous) aquifer properties.

- Complex, time-varying pumping schedules at multiple wells or well fields pumping within a basin.

- Nonlinearities, such as boundary conditions and aquifer properties that change with changes in groundwater levels.

Many of the examples in this report that illustrate various aspects of streamflow depletion are derived from groundwater models of actual stream-aquifer systems. Investigators in those studies chose a numerical-modeling approach, in part because of the complexities listed above.

Groundwater-flow models simulate movement of water from areas of recharge, through an aquifer or an aquifer system, to streams and other features where groundwater discharges. Any groundwater-model program can be used to

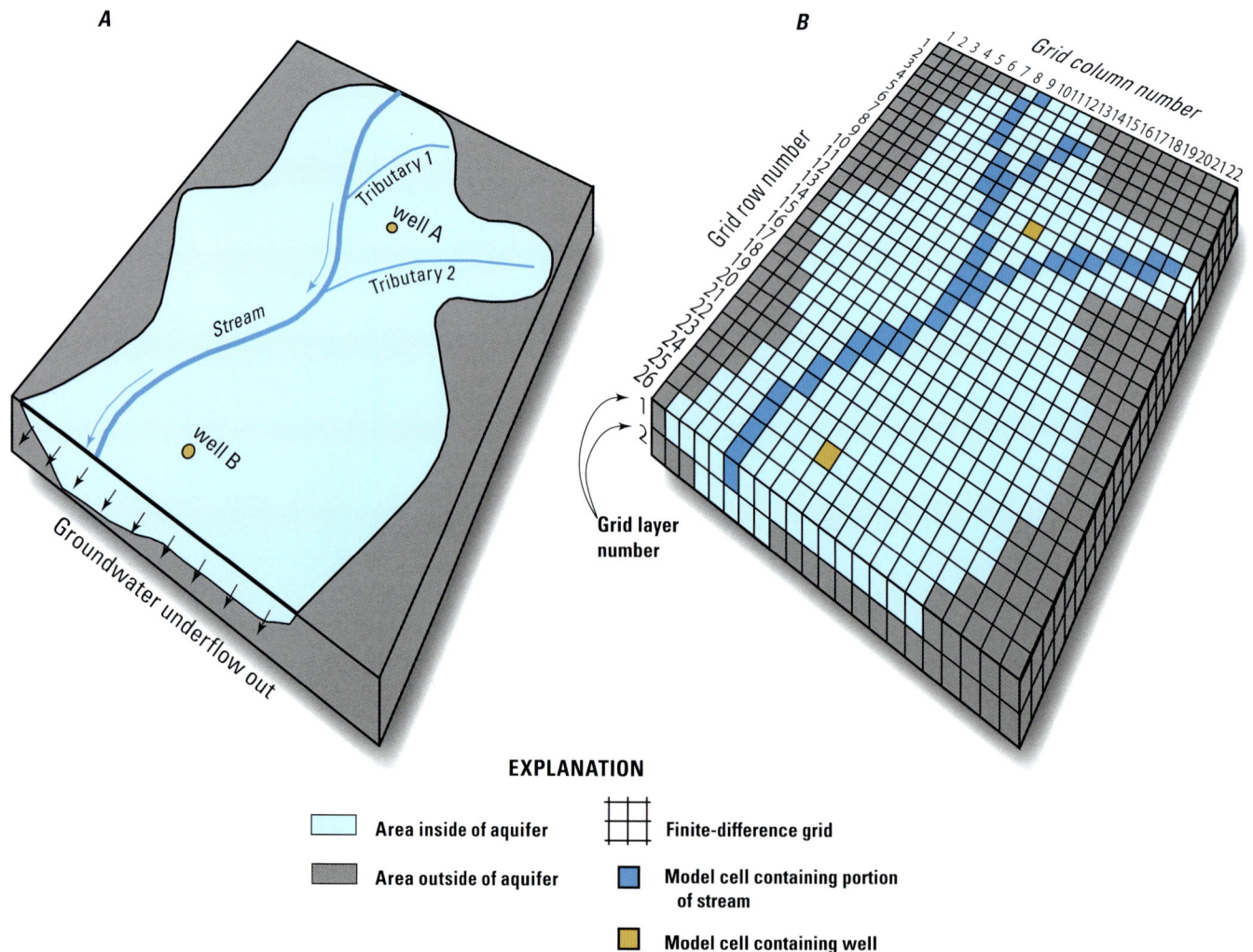

EXPLANATION

- □ Area inside of aquifer
- ▨ Area outside of aquifer
- ⊞ Finite-difference grid
- ■ Model cell containing portion of stream
- ■ Model cell containing well

Figure 42. *A*, Part of a hypothetical stream-aquifer system. *B*, Representation of that system with a finite-difference model grid consisting of 26 rows, 22 columns, and 2 layers of rectangular finite-difference blocks.

simulate depletion, as long as the program carries out rigorous calculations of system water-balance components, including inflow to the aquifer, change in storage within the aquifer, and outflow from the aquifer. The discussion that follows will reference the USGS finite-difference groundwater-model program, MODFLOW (McDonald and Harbaugh, 1988; Harbaugh, 2005), which is used worldwide to simulate many aspects of groundwater flow, including streamflow depletion. This type of model uses a grid of rectangular or square blocks to represent an aquifer (fig. 42*B*). In this example, a portion of a valley-fill aquifer is represented using a finite-difference grid consisting of 26 rows and 22 columns of equally spaced model cells. Aquifer properties are represented as being constant in each grid cell, and locations of boundaries occur either at the center of the cell or along the edges of cells. Use of a regular grid of finite-difference cells leads to approximations of locations of features such as the edges of the aquifer, streams, and wells (fig. 42*B*); however, use of a finer finite-difference grid will allow more accurate representation of locations of these features. In this example, two layers of grid cells were used to represent the aquifer in the vertical dimension.

Steady-State Flow Models

Steady-state groundwater models solve for head (groundwater levels) and flow components for the condition in which inflow rates balance outflow rates, and the rate of storage change in the aquifer is zero. As shown in the example in figure 42, inflow components might include recharge to the aquifer surface (not shown), lateral inflow from the rocks surrounding the aquifer, and flow from some stream segments to the aquifer. Outflow components would include groundwater underflow out of the model domain, flow from the aquifer to stream segments, and discharge by wells. Ultimate effects of pumping on streams (including tributaries) can be computed in three steps as follows:

Step 1. Run the model *without* pumping by a well or wells of interest and record model-computed flow rates to and from stream segments.

Step 2. Run the model *with* pumping by a well or wells of interest and record model-computed flow rates to and from stream segments.

Step 3. Subtract model-computed flow rates in step 2 from corresponding flow rates in step 1 to get net change in flow between the aquifer and streams.

If the pumping cannot increase recharge to the aquifer, or increase lateral inflow, or decrease underflow out, then the total change in flow to and from stream segments will equal the total pumping rate. This type of steady-state analysis cannot address the timing of depletion but is useful in understanding which features would ultimately be affected by the pumping (fig. 43).

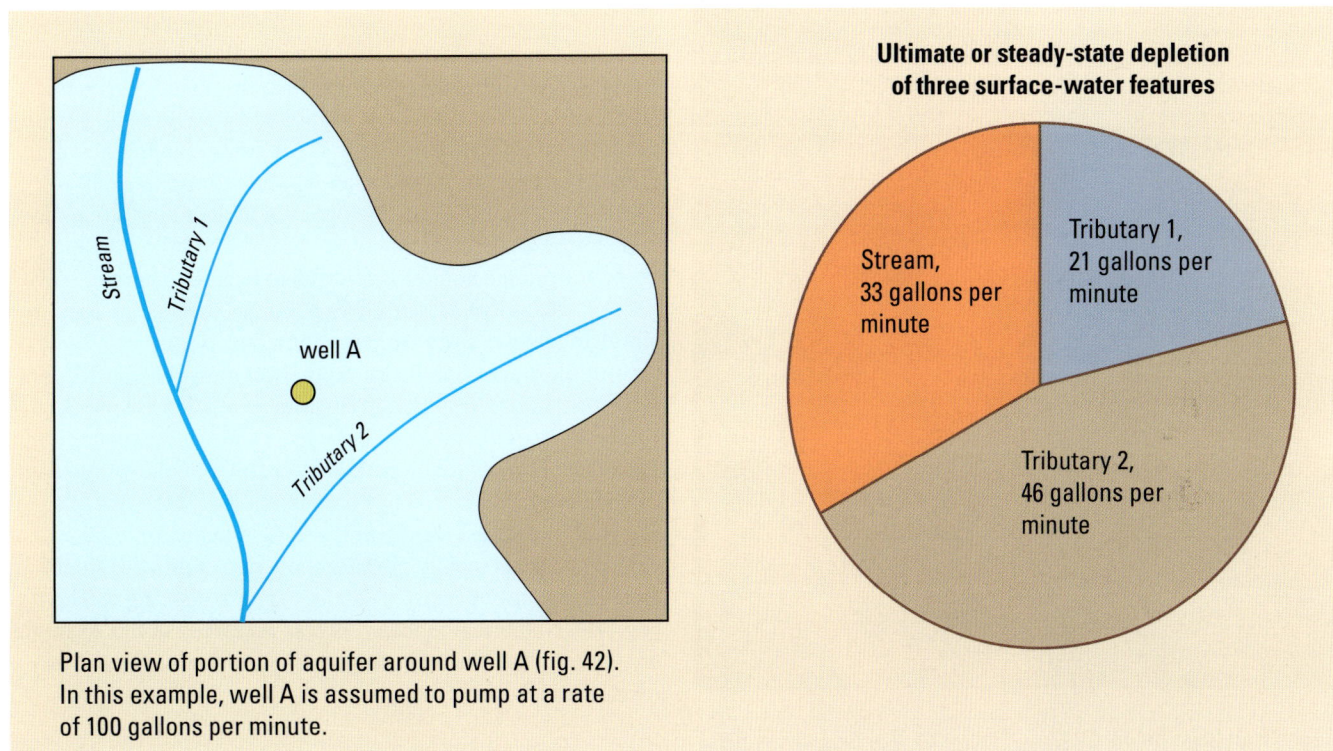

Plan view of portion of aquifer around well A (fig. 42). In this example, well A is assumed to pump at a rate of 100 gallons per minute.

Figure 43. Possible ultimate rate of depletion of different surface-water features by pumping well A at a rate of 100 gallons per minute until steady-state conditions are reached.

Republican River below McCook, Nebraska. The Republican River Compact Administration groundwater model is used to assess groundwater consumptive use in Kansas, Colorado, and Nebraska (*http://www.ksda.gov/ interstate_water_issues/content/142*).

Transient Flow Models

Transient groundwater models solve for head and flow components at discrete intervals of time, called "time steps." In these models, head may change with time and the rate of change in aquifer storage is a component in model water budgets. For the example in figure 42, inflow components to the aquifer would be recharge to the aquifer surface, lateral inflow from surrounding rocks, flow from some stream segments to the aquifer, and the rate that water is released from aquifer storage (the condition that happens when aquifer head declines). Outflow components would include groundwater underflow out of the model domain, flow from the aquifer to stream segments, discharge by wells, and the rate that water is going into aquifer storage (the condition that occurs when aquifer head rises). The latter condition of water going into storage would not occur as a result of pumping, but it is a possible condition in part of the model domain if other water-budget components are varying through time.

The procedure for computing depletion in transient models uses the same three steps outlined above for steady-state models except that these steps must be carried out for each time step for which depletion is to be calculated. For example, if a transient model uses 10 time steps to simulate 1 year of pumping, depletion at a pumping time of 1 year can be calculated by recording flow components at time step 10 in model runs with and without pumping, and computing differences in corresponding components.

Simulated Features that can be Affected by Groundwater Pumping

Although the focus of this report is streamflow depletion, many models simulate additional features including rivers, lakes, springs, wetlands, and evapotranspiration areas. Evaluation of total effects of pumping involves calculating pumping-induced changes in inflow to and outflow from the aquifer from all relevant features. As opposed to the term "streamflow depletion," total change in pumping-induced inflow to and outflow from the aquifer is referred to here as "capture." Table 2 lists select MODFLOW packages that can be used to simulate features from which capture may occur.

In the Upper San Pedro groundwater model (fig. 13), outflow to streams, springs, and riparian vegetation is simulated, respectively, with the Stream, Drain, and Evapotranspiration Packages. For any given pumping location, total capture may include reduced outflow to a combination of these features. For example, at the location of well C in figure 13, total capture consists mostly of streamflow depletion with some evapotranspiration capture and no capture of spring discharge (fig. 44). Numerical models, such as presented in this example, are the only approach to compute capture from different features in a real-world setting.

Table 2. Select MODFLOW packages for representing boundary conditions in which pumping may increase inflow to the aquifer or decrease outflow from the aquifer.

MODFLOW package	Common uses	Possible responses to pumping	Comments
Specified head (CHD)	Well-connected surface-water features	Increased inflow to aquifer, decreased outflow from aquifer	The package sets head in aquifer equal to head in connected surface-water feature
General-head flow (GHB)	Streams, rivers, other surface-water features	Increased inflow to aquifer, decreased outflow from aquifer	A linear boundary condition in which flow between boundary and aquifer is proportional to difference between boundary head and aquifer head
Stream (STR) or Streamflow Routing (SFR)	Streams, rivers	Increased inflow to aquifer, decreased outflow from aquifer	Can calculate stream stage, keeps track of flow in streams, streams may go dry
River (RIV)	Rivers, streams that do not go dry	Increased inflow to aquifer, decreased outflow from aquifer	River stage is specified at each location of cell with a river, seepage rate to aquifer becomes steady if groundwater level drops below bottom of streambed sediments
Drain (DRN)	Agricultural drains, springs	Decreased outflow from aquifer	Discharge to a simulated drain ceases if groundwater level drops below drain altitude
Lake (LAK)	Lakes	Increased inflow to aquifer, decreased outflow from aquifer	Can calculate lake stage, maintains mass balances of lakes, lakes may go dry
Evapotranspiration (EVT)	Groundwater evapotranspiration	Decreased outflow from aquifer	Evapotranspiration ceases if groundwater level drops below a specified level; evapotranspiration is constant with groundwater levels above another specified level

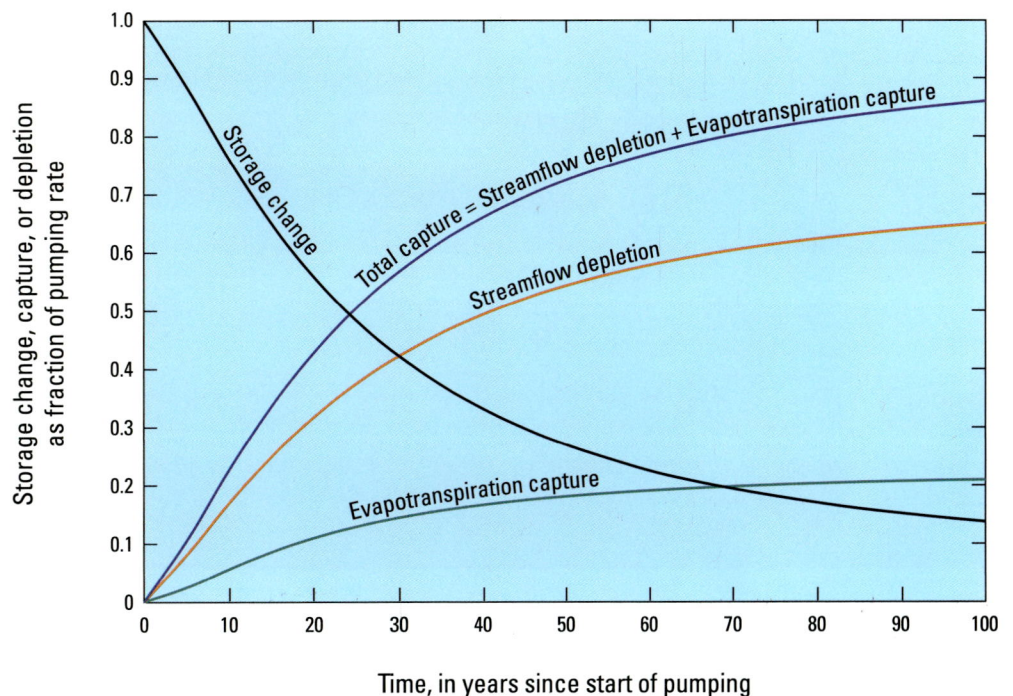

Figure 44. Model-computed streamflow depletion, evapotranspiration capture, and total capture for location of hypothetical well C (see figure 13) in the Upper San Pedro Basin, Arizona.

Photograph by Michael Collier

In addition to depleting streamflow, groundwater pumping can capture groundwater that otherwise would be used by plants (phreatophytes). Riparian trees, shown here, use shallow groundwater along the channel of the Mojave River in California.

Superposition Models

In the suite of methods available for computing depletion, superposition models are an intermediate approach between simple analytical solutions and complex calibrated groundwater-flow models. Unlike flow models, superposition groundwater models do not simulate natural movement of water through an aquifer. Instead of computing head and flow, these models directly compute *change* in head and *change* in flow from an added stress such as pumping. To compute streamflow depletion, the initial pre-pumping state of the superposition model is to have no flow between the stream and the aquifer. After addition of a pumping stress, computed flow from a boundary representing a stream is a direct calculation of total streamflow depletion. Because the natural flow system is not simulated, superposition models cannot determine if the depletion represents reduced groundwater discharge to the stream, increased flow of water from the stream to the aquifer (that is, induced infiltration), or a combination of these two components. Regardless, the streamflow depletion computed by a superposition model is a direct calculation of the reduced availability of surface water in the stream.

Application of the principle of superposition strictly applies to groundwater systems that respond linearly to stresses such as groundwater pumping (Reilly and others, 1987). Linearity of response means that changes from the added stress do not change the aquifer properties or configuration or function of the boundary conditions. Some examples of nonlinear responses include (1) drawdown that causes substantial changes in aquifer saturated thickness and corresponding changes in transmissivity, (2) drawing aquifer water levels below the base of a streambed so that the stream is no longer in direct hydraulic connection with the aquifer, (3) drawing water levels down below the evapotranspiration extinction depth so that evapotranspiration ceases, and (4) drying up a spring or reach of a stream. Many aquifer systems respond linearly to some range of lower stresses, and superposition can be applied in many mildly nonlinear systems (Reilly and others, 1987).

Leake and others (2005) used a superposition modeling approach to compute streamflow depletion from proposed pumping in the C aquifer in northern Arizona (figs. 17–19). In that model, both confined and unconfined areas of the aquifer and complex variations in aquifer thickness were represented. In contrast, Leake, Greer, and others (2008) computed possible depletion of the lower Colorado River using superposition models that were representative of aquifer material of uniform thickness and aquifer properties. In that application, vertical geometry and aquifer properties are treated simplistically as they would be in an analytical solution, yet all complexities of horizontal aquifer and river geometry are represented in greater detail than would be possible by an analytical solution. These types of superposition models can be constructed faster and at less expense than more complex numerical flow models and are useful in gaining an initial understanding of the possible timing of depletion. For details on how to set up a groundwater model to compute changes using superposition, see Reilly and others (1987). Durbin and others (2008) present methods of representing nonlinear boundaries in superposition models.

Simulating the Effects of Other Boundary Conditions on Streamflow Depletion

In addition to boundary conditions representing surface-water features and evapotranspiration (table 2), models can simulate the effects of no-flow or specified-flow boundaries at appropriate locations. For example, the area outside of the aquifer depicted in figure 42 may be crystalline rocks of low permeability. If interchange of water between these rocks and the aquifer is insignificant, the lateral edges of the aquifer shown in the figure could be represented as a no-flow boundary. Alternately, if some mountain-block recharge to the aquifer occurs through these rocks, the interface could be represented as a specified-flow boundary. Whether this boundary is represented as no flow or specified flow, the presence of impermeable or low-permeability rocks tends to speed up the timing of streamflow depletion because drawdown and storage change from pumping cannot extend beyond the boundary.

Ideally, all model boundaries should represent physical features such as the edge of the aquifer or a surface-water boundary. In some cases, it is impractical to construct a model that extends to all physical boundaries. In the example shown in figure 42, the area of interest may be around wells A and B, but the aquifer may extend a great distance down the valley from this area. Using the model domain shown in figure 42, an "artificial" boundary must be implemented to represent flow out of the model domain along model row 26, columns 2–17 in layer 1 and columns 4–8 in layer 2. Options for representing artificial boundaries at this location include (1) specified flow—that is, estimated downvalley flow is input for each boundary cell and the model will compute head at these cells, (2) specified head—that is, head is set to the estimated water level for each boundary cell, and the model will compute flow into or out of the model domain at each of these cells, and (3) head-dependent flow—a boundary head and "conductance" value are specified at each boundary cell so that computed flow into or out of the model varies with changes in head in the connected model cells. No matter which boundary type is selected, proximity of artificial boundaries to pumping wells is a potential problem in calculations of depletion. In figure 42, an artificial boundary along row 26 is distant from well A. Furthermore, well A is surrounded by surface-water boundaries and the natural boundary of the edge of the aquifer. Placement of an artificial boundary in model row 26 is not likely to affect calculations of depletion by pumping well A. In contrast, well B is as close to the artificial boundary as it is to the surface-water boundary. A constant-head artificial boundary along row 26 likely will result in an underestimation of depletion by well B for any given time. In contrast, a specified-flow (including no-flow) artificial boundary at that location would result in an overestimation of the progression of depletion by well B. To calculate depletion for well B, the model should be extended enough distance downstream so that the drawdown from this well does not reach the artificial downstream boundary.

In some aquifers, groundwater divides that approximately underlie watershed boundaries define the extent of a subunit of the aquifer beneath the watershed. If interest is in modeling groundwater processes in the particular watershed, a common practice is to represent the bounding groundwater divides as no-flow boundaries. Under flow conditions that are steady, the groundwater divides are in fact no-flow boundaries because there is no movement of groundwater across the divides. A possible result of added pumping in the watershed, however, is that groundwater divides will be moved outward into adjacent watersheds. Divides that are represented as no-flow boundaries that are fixed in space may result in computed rates of streamflow depletion that occur faster than would be computed using a representation of divides that can move in response to pumping. If drawdown from pumping can propagate to groundwater divides, the best approach is to make the domain of the model large enough so that model boundaries are not on the groundwater divides adjacent to the pumping locations. In the example shown in figure 45, pumping locations A and B are both in watershed 1. Pumping location B is close to the stream segments in watershed 1, and drawdown from pumping at this location probably would not reach the boundaries of the watershed. In this case, a model that includes only the portion of the aquifer underlying watershed 1 may be a reasonable approach to simulating depletion from pumping at location B. In contrast, location A is closer to the watershed boundary than it is to stream segments in watershed 1. Pumping at location A likely would deplete surface water in stream segments in watersheds 1, 2, 3, and 4. Use of a model that includes only watershed 1 for this pumping location would force some of the drawdown, storage change, and depletion that should occur in adjacent watersheds to occur only in watershed 1. The result is an overestimation of depletion in watershed 1 and an underestimation of depletion in adjacent watersheds 2, 3, and 4. To effectively simulate depletion from pumping at location A, the model must include the part of the aquifer underlying watersheds 1, 2, 3, and 4.

Response Functions and Capture Maps

Two important uses of analytical and numerical models are to generate streamflow-depletion response functions and capture maps (which are a type of response function). Response functions characterize the unique functional relation between pumping at a particular location in an aquifer and the resulting depletion in a nearby stream and provide hydrologists and water-resource managers with insight into how a

particular stream or stream reach will respond to pumping at a particular well. Although response functions have been defined and used in different ways (and referred to by different names), all response functions have the common characteristic that they represent a change in streamflow that results from a change in pumping rate at a single well, independently of other pumping or recharge stresses that may be occurring simultaneously within the aquifer[1]. As demonstrated by the many examples provided in this report, the response function for a particular well and streamflow-location pair reflects the combined effects of several factors, including the distance of the well from the stream, the geometry of the aquifer system and stream network, the hydraulic properties of the aquifer and streambed materials, and the vertical depth of pumping from the aquifer.

Theoretically, response functions could be determined by monitoring changes in streamflow that result from pumping at a particular well, but this approach is often not technically feasible because of difficulty in separating depletion changes from streamflow responses to other changes, such as those driven by climate. In practice, response functions are determined by using analytical or numerical models. Model-simulated response functions are shown as either the rate or volume of streamflow depletion that occurs in response to pumping at a particular rate or, alternatively, as dimensionless fractions of the pumping rate or total volume of withdrawal at a well, as described in Box B. Reporting response functions as dimensionless quantities is particularly useful when streamflow depletion responds linearly to pumping, because the dimensionless quantities are constants whose values are independent of the particular pumping rate used for their calculation. For example, if the dimensionless response function were 0.5 for a time and location of interest, the rate of streamflow depletion would be 0.5 Mgal/d for a pumping rate of 1.0 Mgal/d, and 2.0 Mgal/d for a pumping rate of 4.0 Mgal/d. As described previously, a stream-aquifer system is linear if (1) the transmissivity of the aquifer does not change as the pumping rates of the wells change and (2) the rate of flow at

---

[1]Some examples of the application of response functions to stream-aquifer systems include those described by Maddock (1974), Morel-Seytoux and Daly (1975), Morel-Seytoux (1975), Illangasekare and Morel-Seytoux (1982), Danskin and Gorelick (1985), Maddock and Lacher (1991), Reichard (1995), Male and Mueller (1992), Mueller and Male (1993), Fredericks and others (1998), Barlow and others (2003), Cosgrove and Johnson (2004, 2005), and Ahlfeld and Hoque (2008). Although this report focuses on streamflow-depletion response functions, it should be noted that response functions also can be generated for other types of variables that describe the state of a groundwater system, such as groundwater-level declines, groundwater velocities, and aquifer-storage changes (see, for example, Maddock and Lacher, 1991; Gorelick and others, 1993; Ahlfeld and Mulligan, 2000; and Ahlfeld and others, 2005 and 2011).

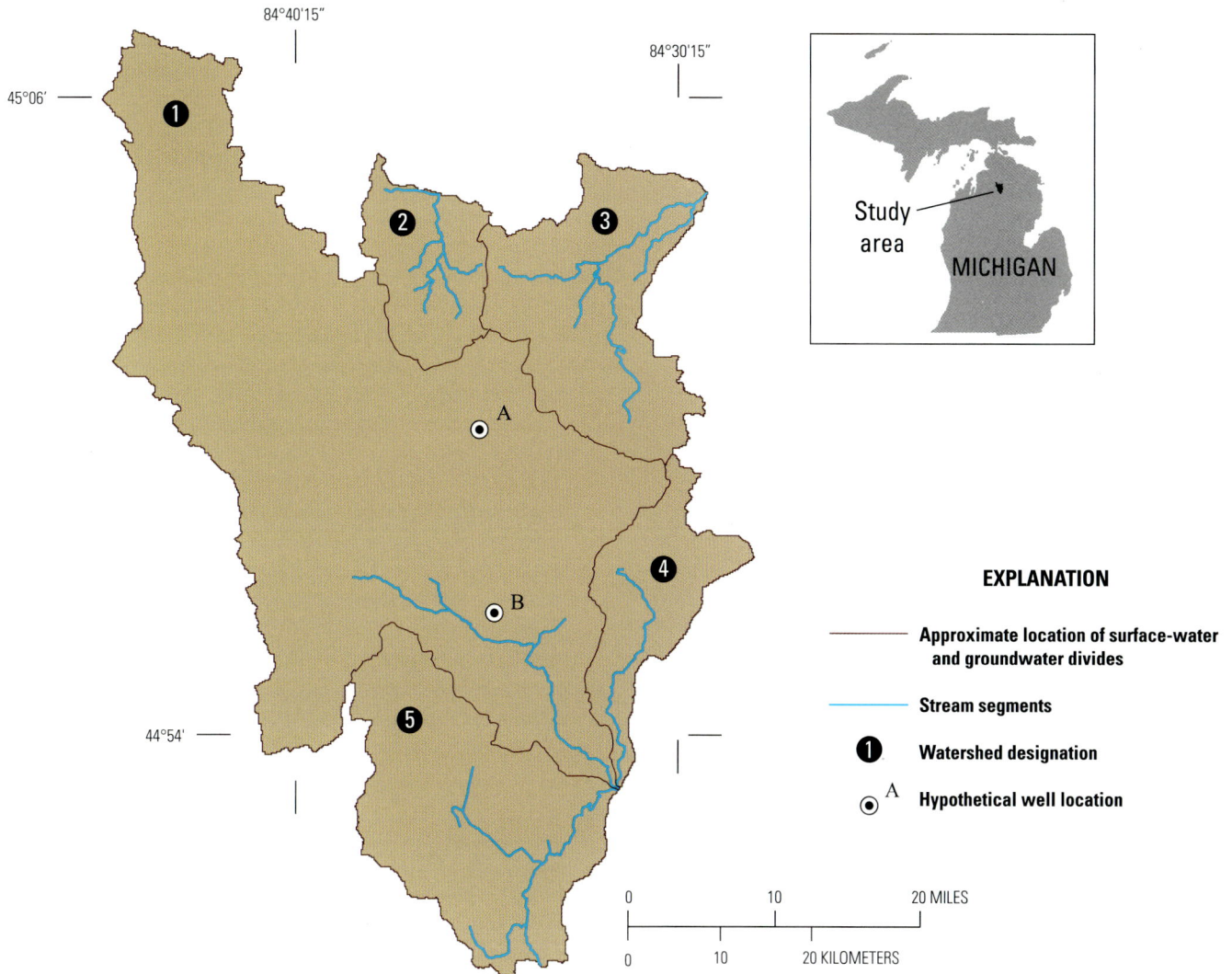

Figure 45.   Five adjacent watersheds in north-central Michigan overlying a groundwater system. Pumping locations A and B are both within watershed 1, but construction of a model to compute depletion for a well at location A will require inclusion of some adjacent watersheds in the model domain (modified from Reeves and others, 2009).

the stream-aquifer boundary is a linear function of the groundwater level near the stream.

Response functions that characterize total depletion of all streams (and sometimes other features) within a basin are referred to here as "global response functions." Conversely, response functions that characterize depletion in a particular stream or segment of a stream are referred to as "local response functions." Furthermore, "transient response functions" characterize depletion through time until some maximum time interval and "steady-state response functions"

characterize ultimate depletion without regard to the time required to reach that state. Some key points relating to these types of response functions are as follows:

1. Transient response functions for each pumping location are defined by a number of values through time.

2. Global transient response functions expressed as a fraction of pumping rate will start at zero at the onset of pumping and will trend toward a maximum value of 1.0, as shown by the curve in figure B–1B in Box B.

3. Local transient response functions may trend toward a value less than 1.0 if the pumping causes depletion in locations in addition to the stream or segment of interest.

4. Steady-state response functions are a single value for each pumping location.

5. Global steady-state response functions are equal to 1.0, assuming that streams are the ultimate source of pumped water.

6. The sum of local steady-state response functions for all stream segments affected by a pumped well is equal to 1.0.

Concepts relating to global and local, transient and steady-state response functions are further illustrated by the two maps in figure 46. Dimensionless response-function values are shown in figure 46*A* for three wells in the watershed after 10 years of pumping. The stream location for which the response coefficients were determined is the outflow point from the basin. For this hypothetical aquifer, the response-function value for well A is largest because the well is closer to the stream network than the other two wells; the value for well C is smallest because it is furthest from the stream network. Figure 46*B* illustrates steady-state local response-function values for one of the tributaries to the main stem (stream segment 1). In this example, the system has reached steady-state conditions, and streamflow depletion is the only source of water to the wells. Each response-function value shown in figure 46*B* represents the change in streamflow at the point just upstream from the confluence of the tributary with the main stem in response to pumping at each of the three wells. The response-function value is largest for well A because it is adjacent to the tributary, whereas the response-function value for well B is lowest because it captures most of its discharge from the main stem (stream segments 3 and 5) and very little discharge from the tributary denoted as stream segment 1.

An alternative approach to calculating response functions for only a few locations is to show maps of the spatial distribution of values of response functions for large regions of an aquifer. Response-function maps are particularly useful for illustrating the effects of pumping location on streamflow depletion within a large set of possible pumping locations within an aquifer (Leake and others, 2010). One approach has been to show values of the global transient response function

for a particular pumping time, such as 10 years (for examples, see Leake, Pool, and Leenhouts, 2008; and Leake and Pool, 2010). Such maps, referred to as "capture maps," provide water-resource managers with a visual tool that can be used to determine the effects of pumping at specific locations on total streamflow depletion. Using values from local transient or steady-state response functions, capture maps also can be created to illustrate effects of pumping location on specific streams or stream segments (Cosgrove and Johnson, 2005; Leake and others, 2010). The goal of any of these types of capture maps is to help convey an understanding of the effects of well placement on depletion in areas of interest and to provide a possible tool for use in siting new wells or recharge facilities.

The procedure for making response-function or capture maps requires use of a well-constructed groundwater model. The model must include streams and other appropriate features as head-dependent boundaries, and any boundaries that do not represent actual physical features must be at distances such that they do not affect calculated depletion. For details on constructing these maps, see Leake and others (2010).

Example capture maps showing global transient response functions for the Upper San Pedro Basin (Leake, Pool, and Leenhouts, 2008) are shown in figure 47. The mapped area is the extent of the lower basin-fill aquifer, represented as layer 4 of the groundwater model by Pool and Dickinson (2007). Global response in this case is mostly from changes to streamflow in the San Pedro and Babocomari Rivers, but also includes minor components of reductions in groundwater evapotranspiration and in springflow (that is, groundwater discharge to springs). For the times shown, 10 years (fig. 47*A*) and 50 years (fig. 47*B*), pumping in the area shaded in the darkest blue indicates that depletion would be between 0 and 10 percent (a fraction from 0 to 0.1) of the pumping rate for that time. Similarly, depletion from pumping in the darkest red area on each map indicates depletion would be between 90 and 100 percent (a fraction from 0.9 to 1.0) of the pumping rate for that time. As would be expected, the general pattern is that depletion from pumping nearer the rivers is greater than from pumping at more distant locations for either time shown; however, amounts of depletion vary along the streams. Leake, Pool, and Leenhouts (2008) attribute the complexities in the patterns shown to spatial variations in aquifer geometry and aquifer properties. A low-permeability clay layer that exists between some pumping locations and connected streams may

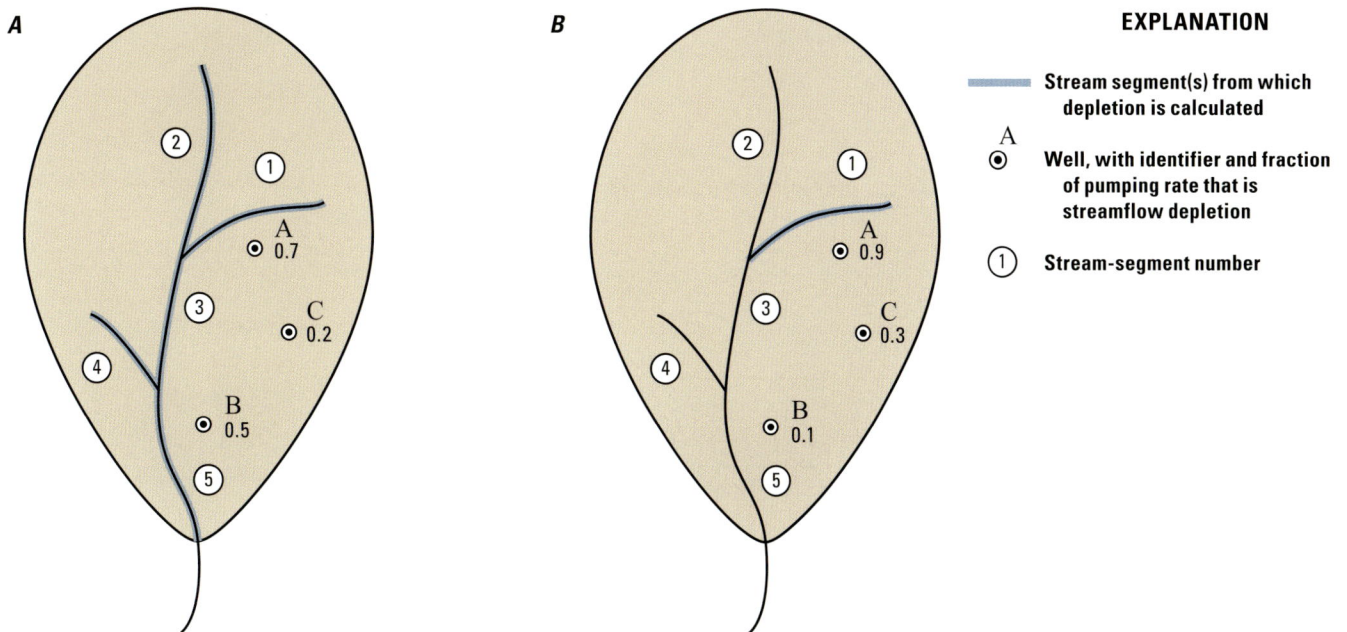

Figure 46.   *A*, Diagram of transient response functions for the outflow point of the basin after 10 years of pumping. *B*, Diagram of steady-state response functions for a tributary stream to the main stem (modified from Leake and others, 2010).

contribute to complexity in the patterns shown. Comparison of the 10-year and 50-year capture maps indicates the progression of depletion through time, with substantially more areas of yellow, orange, and red colors in the 50-year map than in the 10-year map. Because these maps show global response, maps for increased pumping time would be more red, and if pumping time was such that a new steady-state condition would be reached for any pumping location, the map would be solid red.

Other examples of maps to understand depletion as a function of pumping location include mapping of stream-depletion factors by Jenkins and Taylor (1974) and Burns (1983). COHYST Technical Committee (2004), Peterson and others (2008), and Stanton and others (2010) used numerical models to map lines of equal depletion as a fraction of volume pumped at specific times for locations in Nebraska. Some authors have used response-function maps to group wells (or regions of an aquifer) having similar effects on specific

stream reaches into aquifer response zones. Examples of this approach are provided for the Eastern Snake River Plain aquifer in Idaho by Hubbell and others (1997) and Cosgrove and Johnson (2004 and 2005).

In addition to mapping responses for a specific time, it is also possible to construct maps showing the time it would take to reach a particular depletion level of interest. For example, depletion-dominated supply of pumped water (fig. 9) occurs when depletion exceeds half of the pumping rate. Figure 48 shows the time it would take to reach depletion-dominated supply of pumping from the lower basin-fill aquifer in the Upper San Pedro Valley. For most areas adjacent to the Babocomari and San Pedro Rivers, depletion-dominated supply is reached within 20 years of pumping, but in the southern extent of the aquifer and in places along the east and west margins of the aquifer, depletion-dominated supply would not be reached within 100 years (fig. 48).

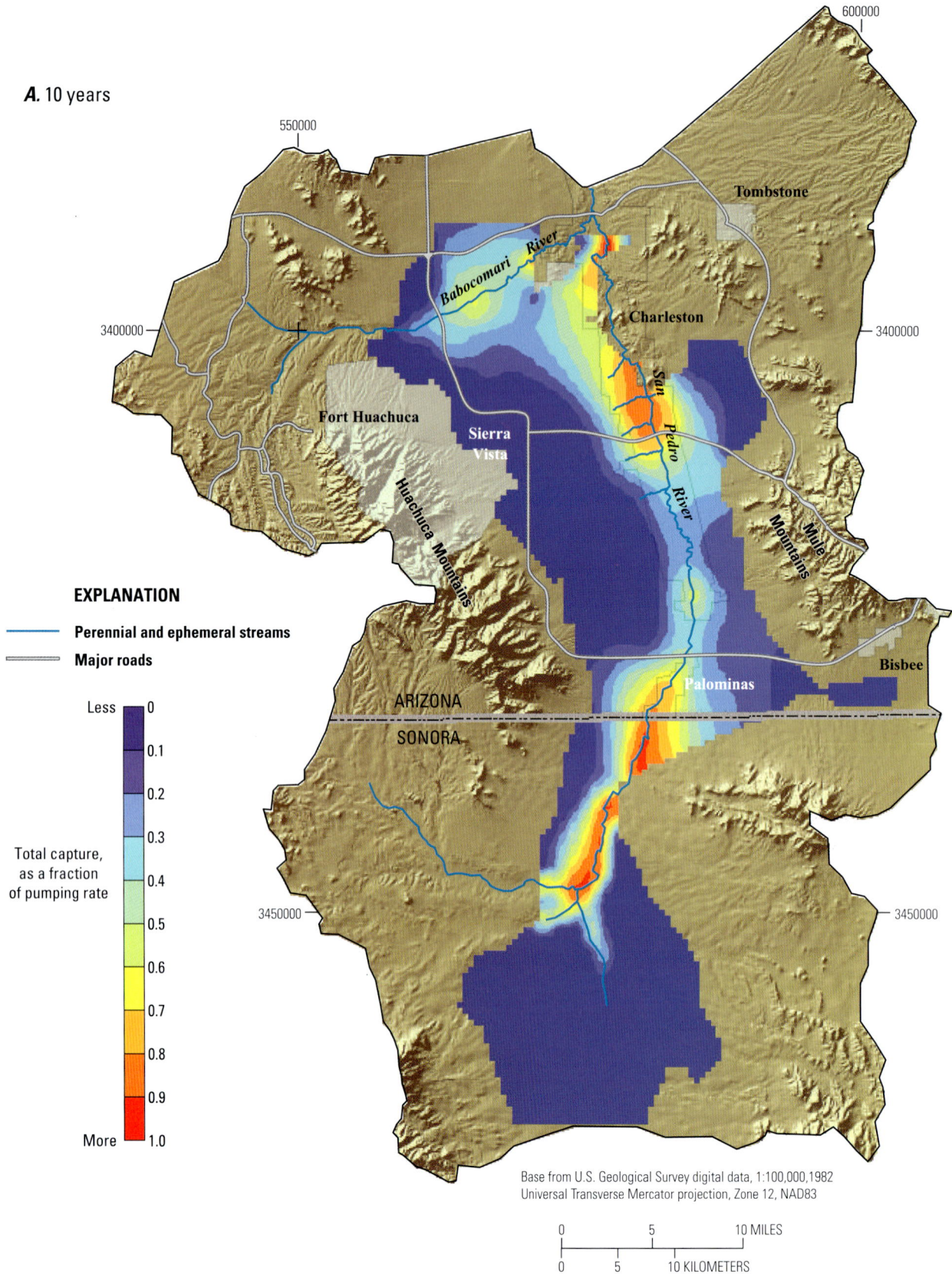

Figure 47. Computed capture of streamflow, riparian evapotranspiration, and springflow that would result for withdrawal of water at a constant rate for, *A,* 10 years and, *B,* 50 years from the lower basin-fill aquifer in the upper San Pedro Basin, Arizona. The color at any location represents the fraction of the withdrawal rate by a well at that location that can be accounted for as changes in outflow from and (or) inflow to the aquifer for model boundaries representing streams, riparian vegetation, and springs (from Leake, Pool, and Leenhouts, 2008).

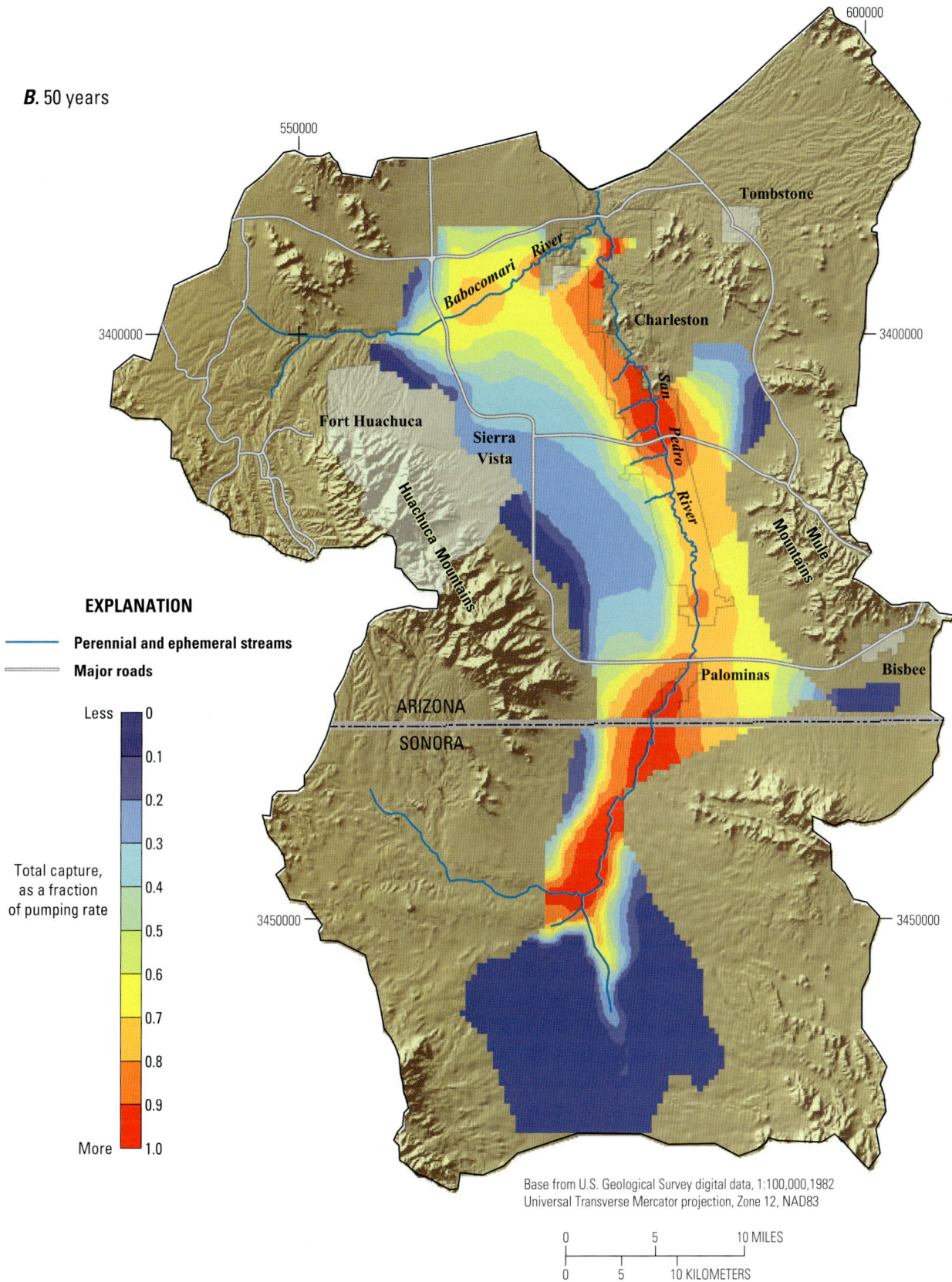

B. 50 years

EXPLANATION

— Perennial and ephemeral streams

═══ Major roads

Total capture, as a fraction of pumping rate

Less — 0
0.1
0.2
0.3
0.4
0.5
0.6
0.7
0.8
0.9
More — 1.0

Base from U.S. Geological Survey digital data, 1:100,000, 1982
Universal Transverse Mercator projection, Zone 12, NAD83

0   5   10 MILES
0   5   10 KILOMETERS

Figure 47. Continued.

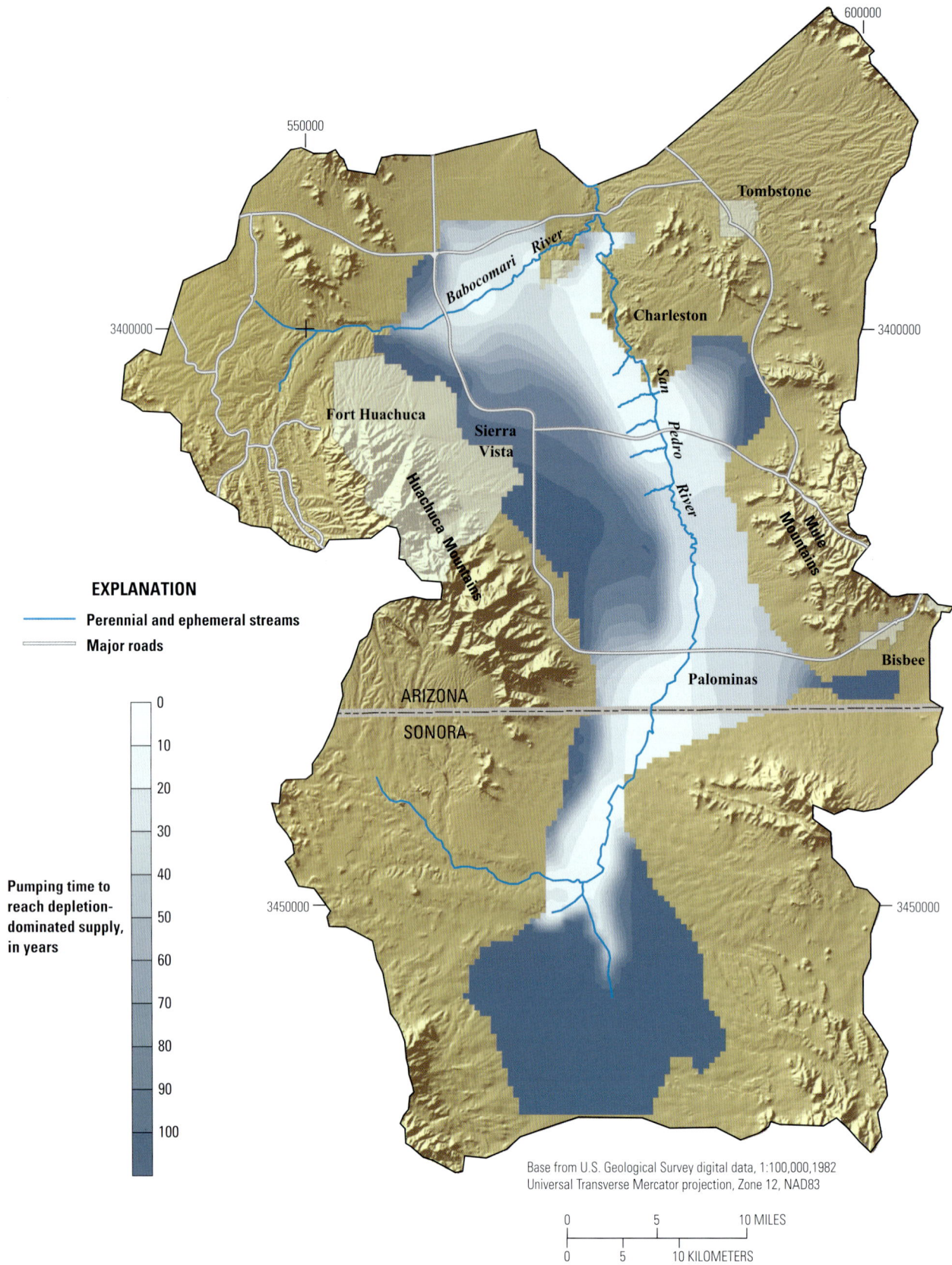

Figure 48.   Computed time to reach a depletion-dominated supply of pumped water for the lower basin-fill aquifer in the Upper San Pedro Basin, Arizona, when streamflow depletion exceeds half of the pumping rate.

Management of Streamflow Depletion

Managing the effects of streamflow depletion by wells is one of the most common and often one of the most challenging aspects of conjunctively managing groundwater and surface-water systems. The effect of a groundwater withdrawal on the timing, rates (or volumes), and locations of streamflow depletions is substantially different from those caused by a surface-water withdrawal, which has an immediate effect on the rate of streamflow at the point of withdrawal. As demonstrated throughout this report, there can be a significant delay between when a well begins to pump and when the impacts of that pumping are realized in nearby streams. These delays can range from days to decades, and in some cases the full impact of pumping may not be realized within a period of time that is meaningful for practical management of a water-supply system. Moreover, unless the pumping site is located very close to the stream, streamflow will not recover immediately after pumping stops because of the residual pumping effects on streamflow depletion. As a result, in many hydrogeologic settings management of pumping rates in response to short-term fluctuations in streamflow conditions such as might be desired during periods of low streamflow or drought is unlikely to have an immediate impact on streamflow (Jenkins, 1968a; Bredehoeft, 2011a).

Other factors, such as determining the locations of streamflow depletions, also complicate management strategies. Streamflow reductions caused by pumping occur both upstream and downstream from the point of withdrawal, and may be distributed among more than one stream; the exact locations of these reductions may not be easily defined without extensive field investigations or modeling studies. Also, many aquifers are tapped by large numbers of wells, and it may not be possible to accurately determine the history of groundwater pumping at each well. It is the sum total of streamflow effects caused by pumping from many wells that need to be managed. A related issue is that an individual well may not produce depletion that is measurable. This is particularly true for large rivers. Finally, aquifers are hidden from view, and even extensive field programs may not be able to define the hydrogeology of a groundwater system in sufficient detail to accurately define the timing of streamflow depletion from an individual well.

In spite of these challenges, water-resource managers often want to understand how pumping rates and pumping schedules might be managed to control the effects of pumping on streamflow depletion. Doing so requires both a long-term perspective (Bredehoeft, 2011a) and an understanding of how streamflow responds to pumping at each well individually and at all wells simultaneously. Several examples of the types of analyses that can be done to determine long-term impacts have been illustrated in this report, such as the generation of response functions and capture maps by use of numerical models. Simulations of specific time-varying and cyclic pumping schedules at individual wells also are useful to determine how aquifer properties and well distance may affect the timing

and variability of streamflow depletion, such as demonstrated for three irrigation wells pumping at various distances from a stream (fig. 21).

An example of some of the issues involved in managing streamflow depletion is illustrated for a typical water-resource management problem, which is to determine pumping schedules that meet water-supply demands while simultaneously meeting minimum streamflow requirements at specific stream locations and for specific periods of time. For this example, an evaluation is made of a single, hypothetical stream that is in hydraulic connection with an aquifer that is pumped from June through August to supply water for irrigation. In the absence of pumping, the annual pattern of streamflow for the hypothetical system ranges from a maximum of 55.0 ft³/s in early spring (March 31) to a minimum of 40.5 ft³/s in early fall (September 30) (fig. 49). Water managers have determined that a minimum streamflow requirement of 35 ft³/s is to be maintained throughout the year to meet instream flow needs. Irrigators want to pump 6 Mgal/d (9.3 ft³/s) from the aquifer from two possible well sites to meet their irrigation requirements. The management problem is to determine whether or not pumping rates can be determined for the two wells to simultaneously meet the irrigation demands and instream-flow requirements.

Because of the simplicity of the physical system, the Glover analytical model is used to determine streamflow depletion caused by different combinations of pumping rates at each well (fig. 49). The first well (A) is located 300 ft from the stream and the second well (B), 1,000 ft from the stream. Three of the many possible combinations of pumping rates at the two wells to meet the irrigation demand are shown in figure 49. When the well closest to the stream is pumped at the full 6 Mgal/d, the minimum streamflow requirement is not met for a short period of time at the end of each pumping cycle (late August into early September). However, when the pumping rate at this well is reduced to 3 Mgal/d and the remaining 3 Mgal/d of the demand is supplied by pumping at the well furthest from the stream, the maximum rate of depletion is reduced and the minimum streamflow requirement is met. The maximum rate of depletion is further reduced as the proportion of pumping from well B increases, with the smallest effect occurring for the case in which all of the withdrawal is from well B. The results shown for this simple stream-aquifer system reflect differences in the underlying streamflow response functions for each well, which in this case result from differences in the distance of each well from the stream.

This simple example demonstrates how pumping rates might be managed to control the timing of streamflow depletion by taking advantage of the variability in streamflow responses to pumping at different wells. For a water-supply system with just a few wells and a single stream location of interest, alternative pumping rates can be tested relatively easily to determine if pumping schedules can be found that simultaneously meet water-supply demands and minimum instream-flow requirements. A trial-and-error testing approach such as this becomes impractical however for a typical hydrogeologic

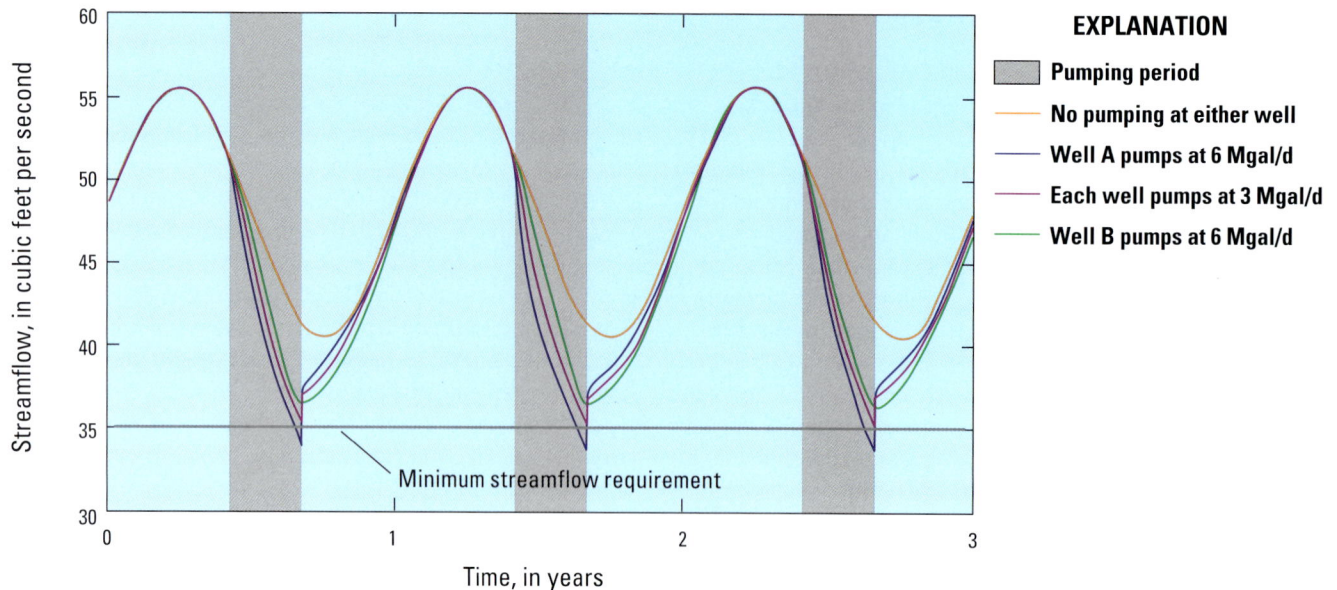

Figure 49.   Streamflow for a hypothetical stream-aquifer system for different pumping conditions. Hydraulic diffusivity of aquifer is 10,000 feet squared per day. Wells are located 300 feet (well A) and 1,000 feet (well B) from the stream. [Rates of streamflow depletion were calculated by using a computer program described in Reeves (2008), which includes the Glover analytical model. The calculated depletion rates were then subtracted from the streamflow hydrograph without pumping (top curve on the figure) to determine the resulting decreased rates of streamflow. Mgal/d, million gallons per day]

setting in which there are multiple pumping wells and multiple streams for which minimum streamflow requirements have been established. For complex settings such as these, a technique called simulation-optimization modeling might be used. In this approach, a numerical simulation model (or, less often, an analytical model) is combined with a mathematical optimization technique to identify pumping schedules that best meet management objectives and constraints. The simulation model accounts for the physical behavior of the stream-aquifer system, whereas the optimization model accounts for the management aspects of the problem. Examples of the use of simulation-optimization modeling for management of streamflow depletion by wells include those described by Young and Bredehoeft (1972), Maddock (1974), Morel-Seytoux and Daly (1975), Morel-Seytoux (1975), Illangasekare and Morel-Seytoux (1982), Bredehoeft and Young (1983), Peralta and others (1988), Matsukawa and others (1992), Male and Mueller (1992), Mueller and Male (1993), Basagaoglu and Marino (1999), Barlow and others (2003), Ahlfeld and Hoque (2008), and Stanton and others (2010). The technique is described in detail by Gorelick and others (1993) and Ahlfeld and Mulligan (2000) and has been implemented for use with some of the widely available groundwater models (for example, the Groundwater-Management Process developed for MODFLOW by Ahlfeld and others, 2005).

An example of the use of simulation-optimization modeling to determine long-term average pumping schedules that meet groundwater-development goals and minimum streamflow requirements is provided by the results of a study for the Big River Basin of Rhode Island by Granato

and Barlow (2005). Minimum streamflow requirements that are protective of aquatic habitats are often not well defined, and, as a result, water-resource and environmental agencies commonly evaluate the effects of alternative streamflow standards on groundwater-development options before implementing a particular regulatory standard. This was the case for the Big River Basin when, at the time of the study, State water-resource and environmental-management agencies were considering more than a dozen alternative minimum streamflow standards for implementation.

A numerical model developed to simulate groundwater flow and groundwater/surface-water interactions within the basin was linked with an optimization model that represented management goals and constraints. The management object was to determine the maximum amount of groundwater that could be pumped from more than a dozen wells in the basin. The maximum rate of withdrawal was limited, however, by constraints placed on the minimum amount of streamflow required at four streamflow locations. Each of the proposed minimum streamflow standards was defined in terms of the minimum streamflow required at each streamflow site per square mile of drainage area to each site. For example, for a defined standard of 0.5 cubic foot per second per square mile [(ft³/s)/mi²], the minimum flow required at a stream location having a 30 mi² drainage area would be 15 ft³/s.

The combined simulation-optimization model was run several times to determine a range of optimal withdrawal rates for alternative definitions of the minimum streamflow standard at the four stream sites. Not surprisingly, the results of the simulation-optimization model indicated that as the minimum streamflow standard was increased, the total amount of pumping within the basin that would be possible decreased (fig. 50). Graphs such as the one shown in figure 50 are often referred to as trade-off curves, because they illustrate the trade-offs that decision makers must consider between minimum streamflow standards and maximum rates of groundwater development. For example, point A on the graph corresponds to a minimum streamflow standard of 0.5 $(ft^3/s)/mi^2$ at each of the four stream sites. For this proposed standard, an average annual pumping rate of 12 Mgal/d from the basin would be possible. Although the overall results of the study could be anticipated without a model—that is, that groundwater development would decrease as the streamflow standard was increased—the specific rates of pumping at each of the wells, and therefore from the basin as a whole, could not. The shape of the curve in figure 50 reflects the unique hydrogeologic and hydrologic conditions within the basin and the distribution of the pumping wells relative to the locations of the streamflow constraint sites.

Both of the examples described in this section and illustrated in figures 49 and 50 were related to managing groundwater withdrawals to meet specified rates of minimum streamflow. However, a number of studies have demonstrated the utility of artificial-recharge strategies at injection wells or artificial-recharge basins to increase streamflow or to offset the effects of withdrawals, such as was illustrated in figure 26. Additional examples of the use of artificial recharge to augment streamflow are provided in the studies by Burns (1984), Bredehoeft and Kendy (2008), and Barber and others (2009).

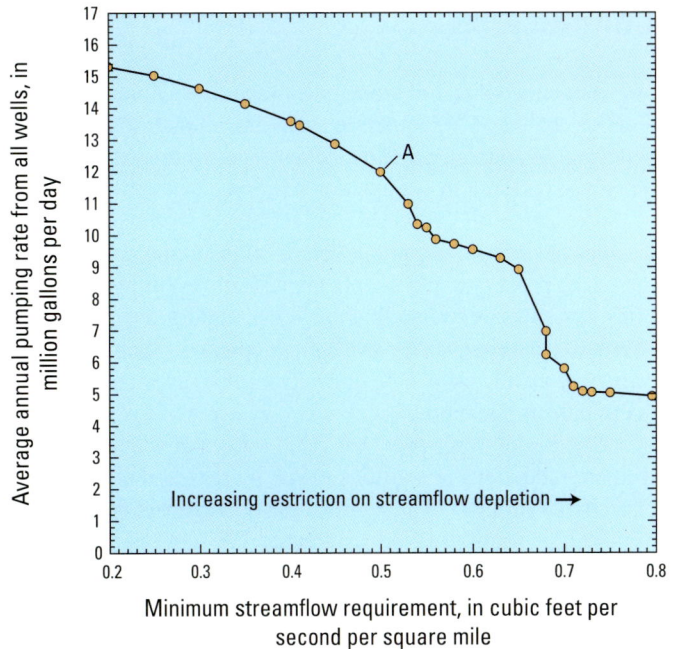

Figure 50. Example application of simulation-optimization modeling to determine trade-offs between minimum streamflow requirements and maximum groundwater pumping rates, Big River Basin, Rhode Island (modified from Granato and Barlow, 2005).

Lateral-move irrigation system used on turf farms, Pawcatuck River Basin, Rhode Island. Concerns about the effects of groundwater and surface-water withdrawals on aquatic habitat in the basin prompted local, State, and Federal agencies to explore water-management strategies that minimize the effects of withdrawals on aquatic habitat (Breault and others, 2009).

Conclusions

Understanding and managing streamflow depletion is a major challenge in regulation and management of groundwater use in coupled groundwater/surface-water systems. Scientific research in conjunction with practical applications of this research to real-world field settings over the past seven decades have made important contributions to the understanding of the processes and factors that affect the timing, locations, and rates of streamflow depletion, and for evaluating alternative approaches for managing depletion. The following primary conclusions can be drawn from this research and the many field applications:

Sources of water to a well: The sources of water to a well are reductions in aquifer storage, increases in the rates of recharge (inflow) to an aquifer, and decreases in the rates of discharge (outflow) from an aquifer. The latter two components are referred to as capture. In many groundwater systems, the primary components of capture are groundwater that would otherwise have discharged to a connected stream or river in the absence of pumping (referred to as captured groundwater discharge) and streamflow drawn into an aquifer because of the pumping (induced infiltration of streamflow).

Components of streamflow depletion: Both captured groundwater discharge and induced infiltration of streamflow result in reductions in the total rate of streamflow. Streamflow depletion, therefore, is the sum of captured groundwater discharge and induced infiltration. Captured groundwater discharge is often the primary component of streamflow depletion, but if pumping rates are relatively large or the locations of withdrawal relatively close to a stream, then induced infiltration may become an important component of streamflow depletion.

Time response of streamflow depletion: Reductions in aquifer storage are the primary source of water to a well during the early stages of pumping. The contribution of water from storage decreases and the contribution from streamflow depletion increases with time as the hydraulic stress caused by pumping expands outward away from the well and reaches one or more areas of the aquifer from which water can be captured. At some point in time, streamflow depletion will be the dominant source of water to the well (that is, more than 50 percent of the discharge from the well) and after an extended period of time may become the only source of water to the well. The time at which streamflow depletion is the only source of water to a well is referred to as the time to full capture.

Factors that affect streamflow depletion: Many factors affect the timing of the response of streamflow depletion to pumping at a particular well. These include the geologic structure, dimensions, and hydraulic properties of the groundwater system; the locations and hydrologic conditions along the boundaries of the groundwater system, including the streams and streambed hydraulic properties; the horizontal and vertical distances of wells from the streams; and pumping schedules at the wells. In a system with predominantly horizontal groundwater flow, well distance and the hydraulic diffusivity of the aquifer are two of the most important factors. Streamflow depletion will occur more rapidly for a well pumping relatively close to a stream from an aquifer having a relatively high value of hydraulic diffusivity and less rapidly for a well pumping far from a stream from an aquifer having a relatively low value of hydraulic diffusivity. In settings in which vertical groundwater-flow components are important, distributions of vertical and horizontal hydraulic conductivity, specific storage, specific yield, and aquifer thickness, in addition to well distance from the stream, are the key properties that control the timing of depletion. Aquifer extent is also an important variable. The time to full capture for wells pumping from narrow river-valley aquifers that are bounded at their margins by relatively impermeable materials can be short (days to years), whereas the time to full capture for wells pumping from regionally extensive aquifer systems can be quite long (years to centuries).

Effects of confining layers on depletion: Various geologic features that act as conduits or barriers to groundwater flow can affect the timing of streamflow depletion from groundwater pumping and also can affect which streams are affected by the pumping. Horizontal or nearly horizontal beds of clay, silt, or other geologic materials that are of substantially lower hydraulic conductivity than adjacent aquifer material may be laterally discontinuous or form laterally extensive confining units that separate adjacent aquifers. Even though confining layers can slow down the progression of depletion in comparison to equivalent aquifer systems without confining layers, it is not reasonable to expect that pumping beneath an extensive confining layer will entirely eliminate depletion. For some well locations, discontinuous confining beds of clay may actually increase the depletion process relative to a condition in which the beds are absent.

Aquifer recharge and streamflow depletion: The long-term average or transient rates of recharge to an aquifer (or the predevelopment rates and directions of flow within an aquifer) will not affect the total amount of depletion that results from pumping a well, because the sources of capture to a well result from changes in the predevelopment recharge and discharge rates to or from an aquifer and not the absolute rates of recharge or discharge themselves. Because the natural rate of recharge does not affect the quantity of streamflow that can be captured by a well, it cannot be assumed that the total amount of groundwater development from an aquifer system is "safe" or "sustainable" at rates up to the long-term average recharge rate. The amount of depletion that can be captured is dependent on the total amount of water in the stream and the amount of reduced streamflow that a community or regulatory authority is willing to accept. However, recharge rates do affect the relative contributions of captured groundwater discharge and induced infiltration to total streamflow depletion: relatively high rates of recharge (or predevelopment flow rates through the aquifer) will result in relatively high rates of captured groundwater discharge, whereas relatively low rates of recharge will result in relatively high rates of induced infiltration.

Distribution of streamflow depletion along stream reaches: Groundwater pumping causes streamflow depletion in streams and stream reaches that are both upgradient and downgradient from the location of withdrawal; the effect of pumping is not confined to those reaches that are immediately adjacent to the well. Some stream reaches will be affected more than others, depending on the distance of the pumped well from each reach and the three-dimensional distribution and hydraulic properties of the sediments that compose the groundwater system and adjoining streambeds. Cumulative streamflow depletion increases in the downstream direction of a basin, and the total amount of depletion in the direction of the outflow point (or points) from the basin will, over time, tend toward the total pumping rate of the well or wells that pump from the basin.

Disconnected and dry stream reaches: Two important assumptions that have been made throughout the report are that the stream and underlying aquifer remain hydraulically connected by a continuous saturated zone and that the stream does not become dry. In extreme cases of large-scale groundwater development and limited streamflow, groundwater levels can be drawn down below the bottom of the streambed and the stream may eventually lose all of its water to the aquifer. Under such conditions, there will not be enough water available from streamflow depletion to offset the pumping by a well or wells in the aquifer.

Streamflow depletion after pumping stops: Streamflow depletion continues after pumping stops because it takes time for groundwater levels to recover from the previous pumping stress and for the depleted aquifer defined by the cone of depression to be refilled with water. The time of maximum streamflow depletion often may occur after pumping has stopped. Eventually, the aquifer and stream may return to their pre-pumping conditions, but the time required for full recovery may be quite long and exceed the total time that the well was pumped. Over the time interval from when pumping starts until the system fully recovers to its prepumping levels, the volume of streamflow depletion will equal the volume of water pumped.

Variable- and cyclic-pumping effects: Pumping schedules at wells fluctuate in response to water-supply demands that change on daily, seasonal, and longer-term intervals. Intermittent- and cyclic-pumping schedules result in variable or cyclic patterns of streamflow depletion, but the overall effect of an aquifer is to damp the variability and amplitude (range) of pumping rates such that the resulting rates of streamflow depletion are less variable and smaller in amplitude than the pumping stress itself. The damping effect is enhanced as the distance of the pumped well increases from a stream or the diffusivity of the aquifer decreases, and at some distance the effects of an intermittent- or cyclic-pumping pattern become indistinguishable from a constant pumping pattern at a cycle (or long-term)-average pumping rate.

Basinwide analyses: Many groundwater basins have hundreds or thousands of pumped wells. Individually, these wells may have little effect on streamflow depletion, but small effects of many wells within a basin can combine to produce substantial effects on streamflow and aquatic habitats. Moreover, basinwide groundwater development typically occurs over a period of several decades, and the resulting cumulative effects on streamflow depletion may not be fully realized for years. As a result of the large number of wells and complex history of development, it is often necessary to take a basinwide perspective to assess the effects of groundwater withdrawals on streamflow depletion.

Streamflow depletion and water quality: Many of the problems associated with streamflow depletion do not require that the two components of depletion—captured groundwater discharge and induced infiltration—be differentiated, or individually quantified. This is the case, for example, for issues that are strictly related to questions of streamflow quantity, such as for water-rights administration or determination of minimum instream-flow requirements for aquatic habitats. For water-quality concerns, however, the relative contribution of captured groundwater discharge and induced infiltration has important implications to the resulting quality of the water in the stream, in the aquifer system, and pumped from wells. As a result, techniques of analysis that are needed to evaluate water-quality problems associated with streamflow depletion must be able to identify the specific components of depletion. For example, analytical solutions and superposition numerical models that can only identify changes in streamflow and not the absolute amount of streamflow will not be appropriate, whereas numerical models, particularly those that can track particles of water through a groundwater system or can simulate solute-transport processes may be.

Field methods for identifying and monitoring streamflow depletion: Two general approaches are used to monitor streamflow depletion: (1) short-term field tests lasting several hours to several months to determine local-scale effects of pumping from a specific well or well field on streams that are in relative close proximity to the location of withdrawal and (2) statistical analyses of hydrologic and climatic data collected over a period of many years to test correlations between long-term changes in streamflow conditions with basinwide development of groundwater resources. Direct measurement of streamflow depletion is made difficult by the limitations of streamflow-measurement techniques to accurately detect a pumping-induced change in streamflow, the ability to differentiate a pumping-induced change in streamflow from other stresses that cause streamflow fluctuations, and by the diffusive effects of a groundwater system that delay the arrival and reduce the peak effect of a particular pumping stress.

Analytical-modeling methods to estimate streamflow depletion: Several analytical solutions to the groundwater-flow equation have been developed to estimate streamflow depletion by wells. These solutions are based on highly simplified representations of field conditions that are necessary to develop mathematical solutions to the groundwater-flow equation but that limit their applicability to real-world field conditions. Some of the important limitations of analytical solutions are that they cannot adequately represent aquifer heterogeneity, the presence of multiple streams or complex stream geometry, or aquifers having complex, three-dimensional geometries. Nevertheless, analytical solutions provide insight into several of the factors that affect streamflow depletion and are often used to make an initial estimate of the effect of a particular well on a nearby stream.

Numerical-modeling methods to estimate streamflow depletion: Numerical models are the most robust method for determining the rates, locations, and timing of streamflow depletion caused by pumping because they are capable of handling many of the common complexities of real groundwater systems. They are the only effective method for determining detailed, basinwide water budgets that account for the effects of complex pumping histories from large numbers of wells on all types of hydrologic features, including streams. Numerical models can be used to generate streamflow-depletion response functions and capture maps. Response functions characterize the unique functional relation between pumping at a particular location and the resulting depletion in a nearby stream or stream network, independently of other pumping or recharge stresses that may be occurring simultaneously within the aquifer. Capture maps, which are a type of response function, show the spatial distribution of response-function values for large regions of an aquifer, and provide a visual tool to illustrate the effects of pumping location on streamflow depletion within a large set of possible pumping locations within an aquifer.

Management of streamflow depletion: Managing streamflow depletion by wells is challenging because of the significant time delays that often occur between when pumping begins and when the effects of that pumping are realized in nearby streams. In many cases, it is not possible to reduce pumping rates during periods of low streamflow to substantially affect flow during the period of stress. Effective management of streamflow depletion requires both a long-term perspective and an understanding of how streamflow responds to pumping at each well individually and at all wells simultaneously. Numerical models are the most effective means to determine the effects of pumping on streamflow and to determine whether or not pumping schedules can be manipulated to meet minimum streamflow requirements. For conditions in which many wells pump from the same basin, the use of numerical models can be enhanced by their coupling with management models that identify the optimal pumping strategies to meet water-resource goals and constraints.

Depletion of other hydrologic features: Most aquifer systems are complex, with water moving from areas of recharge through geologic materials and discharging to streams, springs, rivers, and wetlands, and by plants that use groundwater. The introduction of groundwater pumping can affect all features connected to an aquifer. The emphasis of this report has been on the effects of pumping on connected streams, although most of the discussion that has been presented is equally applicable to other connected features.

Acknowledgments

The authors thank William Alley, Marshall Gannett, Thomas Reilly, and Kay Hedrick-Naugle of the U.S. Geological Survey for their helpful technical and editorial comments on earlier drafts of this report and Christine Mendelsohn, U.S. Geological Survey, for preparation of the illustrations and layout of the final report. The authors also thank Michael Collier, Bob Herrmann, and U.S. Geological Survey personnel for allowing the use of their photographs in this report.

References Cited

Ahlfeld, D.P., Barlow, P.M., and Baker, K.M., 2011, Documentation for the State Variables Package for the Groundwater-Management Process of MODFLOW–2005 (GWM–2005): U.S. Geological Survey Techniques and Methods 6–A36, 45 p.

Ahlfeld, D.P., Barlow, P.M., and Mulligan, A.E., 2005, GWM—A Ground-Water Management Process for the U.S. Geological Survey modular ground-water model (MODFLOW–2000): U.S. Geological Survey Open-File Report 2005–1072, 124 p.

Ahlfeld, D.P., and Hoque, Yamen, 2008, Impact of simulation model solver performance on groundwater management problems: Ground Water, v. 46, no. 5, p. 716–726.

Ahlfeld, D.P., and Mulligan, A.E., 2000, Optimal management of flow in groundwater systems: San Diego, Calif., Academic Press, 185 p.

Alley, W.M., Healy, R.W., LaBaugh, J.W., and Reilly, T.E., 2002, Flow and storage in groundwater systems: Science, v. 296, p. 1985–1990.

Alley, W.M., and Leake, S.A., 2004, The journey from safe yield to sustainability: Ground Water, v. 42, no. 1, p. 12–16.

Alley, W.M., Reilly, T.E., and Franke, O.L., 1999, Sustainability of ground-water resources: U.S. Geological Survey Circular 1186, 79 p.

Anderson, M.P., 2005, Heat as a ground water tracer: Ground Water, v. 43, no. 6, p. 951–968.

Armstrong, D.S., Richards, T.A., and Parker, G.W., 2001, Assessment of habitat, fish communities, and streamflow requirements for habitat protection, Ipswich River, Massachusetts, 1998–99: U.S. Geological Survey Water-Resources Investigations Report 01–4161, 72 p.

Barber, M.E., Hossain, Akram, Covert, J.J., and Gregory, G.J., 2009, Augmentation of seasonal low stream flows by artificial recharge in the Spokane Valley–Rathdrum Prairie aquifer of Idaho and Washington, U.S.A.: Hydrogeology Journal, v. 17, p. 1459–1470.

Barlow, J.R.B., and Clark, B.R., 2011, Simulation of water-use conservation scenarios for the Mississippi Delta using an existing regional groundwater flow model: U.S. Geological Survey Scientific Investigations Report 2011–5019, 14 p.

Barlow, P.M., 1997, Dynamic models for conjunctive management of stream-aquifer systems of the glaciated Northeast: Storrs, Conn., University of Connecticut, Ph.D. dissertation, 256 p.

Barlow, P.M., 2000, Documentation of computer program STRMDEPL—A program to calculate streamflow depletion by wells using analytical solutions, in Zarriello, P.J., and Ries, K.G., III, A precipitation-runoff model for analysis of the effects of water withdrawals on streamflow, Ipswich River Basin, Massachusetts: U.S. Geological Survey Water-Resources Investigations Report 00–4029, p. 75–89.

Barlow, P.M., Ahlfeld, D.P., and Dickerman, D.C., 2003, Conjunctive-management models for sustained yield of stream-aquifer systems: Journal of Water Resources Planning and Management, v. 129, no. 1, p. 35–48.

Barlow, P.M., and Dickerman, D.C., 2001, Numerical-simulation and conjunctive-management models of the Hunt-Annaquatucket-Pettaquamscutt stream-aquifer system, Rhode Island: U.S. Geological Survey Professional Paper 1636, 88 p.

Baron, J.S., Poff, N.L., Angermeier, P.L., Dahm, C.N., Gleick, P.H., Hairston, N.G., Jr., Jackson, R.B., Johnston, C.A., Richter, B.D., and Steinman, A.D., 2002, Meeting ecological and societal needs for freshwater: Ecological Applications, v. 12, no. 5, p. 1247–1260.

Basagaoglu, Hakan, and Marino, M.A., 1999, Joint management of surface and ground water supplies: Ground Water, v. 37, no. 2, p. 214–222.

Bourg, A.C.M., and Bertin, Clotilde, 1993, Biogeochemical processes during the infiltration of river water into an alluvial aquifer: Environmental Science and Technology, v. 27, no. 4, p. 661–666.

Breault, R.F., Zarriello, P.J., Bent, G.C., Masterson, J.P., Granato, G.E., Scherer, J.E., and Crawley, K.M., 2009, Effects of water-management strategies on water resources in the Pawcatuck River Basin, southwestern Rhode Island and southeastern Connecticut: U.S. Geological Survey Circular 1340, 16 p.

Bredehoeft, J., and Durbin, T., 2009, Ground water development—The time to full capture problem: Ground Water, v. 47, no. 4, p. 506–514.

Bredehoeft, J.D., 2002, The water budget myth revisited—Why hydrogeologists model: Ground Water, v. 40, no. 4, p. 340–345.

Bredehoeft, J.D., 2011a, Hydrologic trade-offs in conjunctive use management: Ground Water, v. 49, no. 4, p. 468–475.

Bredehoeft, J.D., 2011b, Monitoring regional groundwater extraction—The problem: Ground Water, v. 49, no. 6, p. 808–814.

Bredehoeft, J.D., Papadopulos, S.S., and Cooper, H.H., Jr., 1982, Groundwater—The water-budget myth, in Scientific basis of water-resource management: Washington, D.C., National Academy Press, p. 51–57.

Bredehoeft, J.D., and Young, R.A., 1983, Conjunctive use of groundwater and surface water for irrigated agriculture—Risk aversion: Water Resources Research, v. 19, no. 5, p. 1111–1121.

Bredehoeft, John, and Kendy, Eloise, 2008, Strategies for offsetting seasonal impacts of pumping on a nearby stream: Ground Water, v. 46, no. 1, p. 23–29.

Brunner, Philip, Cook, P.G., and Simmons, C.T., 2011, Disconnected surface water and groundwater—From theory to practice: Ground Water, v. 49, no. 4, p. 460–467.

Burns, A.W., 1983, Simulated hydrologic effects of possible ground-water and surface-water management alternatives in and near the Platte River, South-Central, Nebraska: U.S. Geological Survey Professional Paper 1277–G, 30 p.

Burns, A.W., 1984, Simulated effects of an artificial-recharge experiment near Proctor, Logan County, Colorado: U.S. Geological Survey Water-Resources Investigations Report 84–4010, 17 p.

Butler, J.J., Jr., Zhan, Xiaoyong, and Zlotnik, V.A., 2007, Pumping-induced drawdown and stream depletion in a leaky aquifer system: Ground Water, v. 45, no. 2, p. 178–186.

Butler, J.J., Jr., Zlotnik, V.A., and Tsou, Ming-Shu, 2001, Drawdown and stream depletion produced by pumping in the vicinity of a partially penetrating stream: Ground Water, v. 39, no. 5, p. 651–659.

Chen, Xunhong, 2001, Migration of induced-infiltration stream water into nearby aquifers due to seasonal ground water withdrawal: Ground Water, v. 39, no. 5, p. 721–728.

Chen, Xunhong, and Chen, Xi, 2003, Effects of aquifer anisotropy on the migration of infiltrated stream water to a pumping well: Journal of Hydrologic Engineering, v. 8, no. 5, p. 287–293.

Chen, Xunhong, and Shu, Longcang, 2002, Stream-aquifer interactions—Evaluation of depletion volume and residual effects from ground water pumping: Ground Water, v. 40, no. 3, p. 284–290.

Chen, Xunhong, and Yin, Yanfeng, 2001, Streamflow depletion—Modeling of reduced baseflow and induced stream infiltration from seasonally pumped wells: Journal of the American Water Resources Association, v. 37, no. 1, p. 185–195.

Chen, Xunhong, and Yin, Yanfeng, 2004, Semianalytical solutions for stream depletion in partially penetrating streams: Ground Water, v. 42, no. 1, p. 92–96.

Christensen, Steen, 2000, On the estimation of stream flow depletion parameters by drawdown analysis: Ground Water, v. 38, no. 5, p. 726–734.

COHYST Technical Committee, 2004, The 40-year, 28-percent stream depletion lines for the COHYST area west of Elm Creek, Nebraska: Platte River Cooperative Hydrology Study (COHYST), 12 p., accessed October 3, 2011, at *http://cohyst.dnr.ne.gov/adobe/dc012_28-40_lines_092104.pdf.*

Conrad, L.P., and Beljin, M.S., 1996, Evaluation of an induced infiltration model as applied to glacial aquifer systems: Water Resources Bulletin, v. 32, no. 6, p. 1209–1220.

Constantz, Jim, 2008, Heat as a tracer to determine stream-bed water exchanges: Water Resources Research, v. 44, W00D10, 20 p. (Also available at *http://dx.doi.org/10.1029/2008WR006996.*)

Cosgrove, D.M., and Johnson, G.S., 2004, Transient response functions for conjunctive water management in the Snake River Plain, Idaho: Journal of the American Water Resources Association, v. 40, no. 6, p. 1469–1482.

Cosgrove, D.M., and Johnson, G.S., 2005, Aquifer management zones based on simulated surface-water response functions: Journal of Water Resources Planning and Management, v. 131, no. 2, p. 89–100.

Danskin, W.R., and Gorelick, S.M., 1985, A policy evaluation tool—Management of a multiaquifer system using controlled stream recharge: Water Resources Research, v. 21, no. 11, p. 1731–1747.

Darama, Yakup, 2001, An analytical solution for stream depletion by cyclic pumping of wells near streams with semipervious beds: Ground Water, v. 39, no. 1, p. 79–86.

Domenico, P.A., and Schwartz, F.W., 1990, Physical and chemical hydrogeology: New York, John Wiley and Sons, 824 p.

Dudley, R.W., and Stewart, G.J., 2006, Estimated effects of ground-water withdrawals on streamwater levels of the Pleasant River near Crebo Flats, Maine, July 1 to September 30, 2005: U.S. Geological Survey Scientific Investigations Report 2006–5268, 14 p.

Durbin, Timothy, Delemos, David, and Rajagopal-Durbin, Aparna, 2008, Application of superposition with nonlinear head-dependent fluxes: Ground Water, v. 46, no. 2, p. 251–258.

Farnsworth, C.E., and Hering, J.G., 2011, Inorganic geochemistry and redox dynamics in bank filtration settings: Environmental Science and Technology, v. 45, p. 5079–5087.

Fetter, C.W., 2001, Applied hydrogeology (4th ed.): Upper Saddle River, N.J., Prentice Hall, 598 p.

Fleckenstein, Jan, Anderson, Michael, Fogg, Graham, and Mount, Jeffrey, 2004, Managing surface water-groundwater to restore fall flows in the Cosumnes River: Journal of Water Resources Planning and Management, v. 130, no. 4, p. 301–310.

Fox, G.A., 2004, Evaluation of a stream aquifer analysis test using analytical solutions and field data: Journal of the American Water Resources Association, v. 40, no. 3, p. 755–763.

Fox, G.A., DuChateau, P., and Durnford, D.S., 2002, Analytical model for aquifer response incorporating distributed stream leakage: Ground Water, v. 40, no. 4, p. 378–384.

Fredericks, J.W., Labadie, J.W., and Altenhofen, J.M., 1998, Decision support system for conjunctive stream-aquifer management: Journal of Water Resources Planning and Management, v. 124, no. 2, p. 69–78.

Freeze, R.A., and Cherry, J.A., 1979, Groundwater: Englewood Cliffs, N.J., Prentice Hall, 604 p.

Gannett, M.W., and Lite, K.E., Jr., 2004, Simulation of regional ground-water flow in the upper Deschutes Basin, Oregon: U.S. Geological Survey Water-Resources Investigations Report 03–4195, 86 p.

Gleeson, Tom, Alley, W.M., Allen, D.M., Sophocleous, M.A., Zhou, Yangxiao, Taniguchi, Makoto, and VanderSteen, Jonathan, 2012, Towards sustainable groundwater use—Setting long-term goals, backcasting, and managing adaptively: Ground Water, v. 50, no. 1, p. 19–26.

Glennon, Robert, 2002, Water follies—Groundwater pumping and the fate of America's fresh waters: Washington, D.C., Island Press, 314 p.

Glover, R.E., 1974, Transient ground water hydraulics: Littleton, Colo., Water Resources Publications, 413 p.

Glover, R.E., and Balmer, G.G., 1954, River depletion resulting from pumping a well near a river: Transactions of the American Geophysical Union, v. 35, no. 3, p. 468–470.

Gorelick, S.M., Freeze, R.A., Donohue, David, and Keely, J.F., 1993, Groundwater contamination—Optimal capture and containment: Chelsea, Mich., Lewis Publishers, 385 p.

Granato, G.E., and Barlow, P.M., 2005, Effects of alternative instream-flow criteria and water-supply demands on ground-water development options in the Big River Area, Rhode Island: U.S. Geological Survey Scientific Investigations Report 2004–5301, 110 p.

Hantush, M.S., 1965, Wells near streams with semipervious beds: Journal of Geophysical Research, v. 70, no. 12, p. 2829–2838.

Harbaugh, A.W., 2005, MODFLOW–2005, The U.S. Geological Survey modular ground-water model—The Ground-Water Flow Process: U.S. Geological Survey Techniques and Methods 6–A16 [variously paged].

Hayashi, Masaki, and Rosenberry, D.O., 2002, Effects of ground water exchange on the hydrology and ecology of surface water: Ground Water, v. 40, no. 3, p. 309–316.

Heath, R.C., 1983, Basic ground-water hydrology: U.S. Geological Survey Water-Supply Paper 2220, 84 p.

Hubbell, J.M., Bishop, C.W., Johnson, G.S., and Lucas, J.G., 1997, Numerical ground-water flow modeling of the Snake River Plain aquifer using the superposition technique: Ground Water, v. 35, no. 1, p. 59–66.

Hunt, Bruce, 1999, Unsteady stream depletion from ground water pumping: Ground Water, v. 37, no. 1, p. 98–102.

Hunt, Bruce, 2003a, Unsteady stream depletion when pumping from semiconfined aquifer: Journal of Hydrologic Engineering, v. 8, no. 1, p. 12–19.

Hunt, Bruce, 2003b, Field-data analysis for stream depletion: Journal of Hydrologic Engineering, v. 8, no. 4, p. 222–225.

Hunt, Bruce, 2008, Stream depletion for streams and aquifers with finite widths: Journal of Hydrologic Engineering, v. 13, no. 2, p. 80–89.

Hunt, Bruce, Weir, Julian, and Clausen, Bente, 2001, A stream depletion field experiment: Ground Water, v. 39, no. 2, p. 283–289.

Hurr, R.T., Schneider, P.A., Jr., and others, 1972, Hydrogeologic characteristics of the valley-fill aquifer in the Julesburg reach of the South Platte River Valley, Colorado: U.S. Geological Survey Open-File Report 73–125, 2 p., 6 pls.

Illangasekare, Tissa, and Morel-Seytoux, H.J., 1982, Stream-aquifer influence coefficients as tools for simulation and management: Water Resources Research, v. 18, no. 1, p. 168–176.

Jenkins, C.T., 1968a, Computation of rate and volume of stream depletion by wells: U.S. Geological Survey Techniques of Water-Resources Investigations, book 4, chap. D1, 17 p.

Jenkins, C.T., 1968b, Techniques for computing rate and volume of stream depletion by wells: Ground Water, v. 6, no. 2, p. 37–46.

Jenkins, C.T., 1968c, Electric-analog and digital-computer model analysis of stream depletion by wells: Ground Water, v. 6, no. 6, p. 27–34.

Jenkins, C.T., and Taylor, O.J., 1974, A special planning technique for stream-aquifer systems: U.S. Geological Survey Open-File Report 74–242, 16 p.

Kelly, B.P., 2002, Ground-water flow simulation and chemical and isotopic mixing equation analysis to determine source contributions to the Missouri River alluvial aquifer in the vicinity of the Independence, Missouri, well field: U.S. Geological Survey Water-Resources Investigations Report 02–4208, 53 p.

Kelly, B.P., and Rydlund, P.H., Jr., 2006, Water-quality changes caused by riverbank filtration between the Missouri River and three pumping wells of the Independence, Missouri, well field, 2003–05: U.S. Geological Survey Scientific Investigations Report 2006–5174, 48 p.

Kendy, Eloise, and Bredehoeft, J.D., 2006, Transient effects of groundwater pumping and surface-water-irrigation returns on streamflow: Water Resources Research, v. 42, W08415, 11 p. (Also available at *http://dx.doi.org/10.1029/2005WR004792*).

Kenny, J.F., Barber, N.L., Hutson, S.S, Linsey, K.S., Lovelace, J.K., and Maupin, M.A., 2009, Estimated use of water in the United States in 2005: U.S. Geological Survey Circular 1344, 52 p.

Kollet, S.J., and Zlotnik, V.A., 2003, Stream depletion predictions using pumping test data from a heterogeneous stream-aquifer system (a case study from the Great Plains, U.S.A.): Journal of Hydrology, v. 281, no. 1, p. 96–114.

Konikow, L.F., Goode, D.J., and Hornberger, G.Z., 1996, A three-dimensional method-of-characteristics solute-transport model (MOC3D): U.S. Geological Survey Water-Resources Investigations Report 96–4267, 87 p.

Leake, S.A., 2011, Capture—Rates and directions of ground-water flow don't matter!: Ground Water, v. 49, no. 4, p. 456–458.

Leake, S.A., Greer, William, Watt, Dennis, and Weghorst, Paul, 2008, Use of superposition models to simulate possible depletion of Colorado River water by ground-water withdrawal: U.S. Geological Survey Scientific Investigations Report 2008–5189, 25 p.

Leake, S.A., and Haney, Jeanmarie, 2010, Possible effects of groundwater pumping on surface water in the Verde Valley, Arizona: U.S. Geological Survey Fact Sheet 2010–3108, 4 p.

Leake, S.A., Hoffmann, J.P., and Dickinson, J.E., 2005, Numerical ground-water change model of the C aquifer and effects of ground-water withdrawals on stream depletion in selected reaches of Clear Creek, Chevelon Creek, and the Little Colorado River, Northeastern Arizona: U.S. Geological Survey Scientific Investigations Report 2005–5277, 29 p.

Leake, S.A., and Pool, D.R., 2010, Simulated effects of groundwater pumping and artificial recharge on surface-water resources and riparian vegetation in the Verde Valley sub-basin, central Arizona: U.S. Geological Survey Scientific Investigations Report 2010–5147, 18 p.

Leake, S.A., Pool, D.R., and Leenhouts, J.M., 2008, Simulated effects of ground-water withdrawals and artificial recharge on discharge to streams, springs, and riparian vegetation in the Sierra Vista Subwatershed of the Upper San Pedro Basin, southeastern Arizona: U.S. Geological Survey Scientific Investigations Report 2008–5207, 14 p.

Leake, S.A., Reeves, H.W., and Dickinson, J.E., 2010, A new capture fraction method to map how pumpage affects surface water flow: Ground Water, v. 48, no. 5, p. 690–700.

Linsley, R.K., Jr., Kohler, M.A., and Paulhus, J.L.H., 1982, Hydrology for engineers (3d ed.): New York, McGraw-Hill, 508 p.

Lohman, S.W., 1972, Ground-water hydraulics: U.S. Geological Survey Professional Paper 708, 70 p.

Lough, H.K., and Hunt, Bruce, 2006, Pumping test evaluation of stream depletion parameters: Ground Water, v. 44, no. 4, p. 540–546.

Macler, B.A., 1995, Developing a national drinking water regulation for disinfection of ground water: Ground Water Monitoring and Remediation, v. 15, no. 4, p. 77–84.

Maddock, Thomas, III, 1974, The operation of a stream-aquifer system under stochastic demands: Water Resources Research, v. 10, no. 1, p. 1–10.

Maddock, Thomas, III, and Lacher, L.J., 1991, Drawdown, velocity, storage, and capture response functions for multiaquifer systems: Water Resources Research, v. 27, no. 11, p. 2885–2898.

Maddock, Thomas, III, and Vionnet, L.B., 1998, Groundwater capture processes under a seasonal variation in natural recharge and discharge: Hydrogeology Journal, v. 6, no. 1, p. 24–32.

Male, J.W., and Mueller, F.A., 1992, Model for prescribing ground-water use permits: Journal of Water Resources Planning and Management, v. 118, no. 5, p. 543–561.

Matsukawa, Joy, Finney, B.A., and Willis, Robert, 1992, Conjunctive-use planning in Mad River Basin, California: Journal of Water Resources Planning and Management, v. 118, no. 2, p. 115–132.

McCarthy, K.A., McFarland, W.D., Wilkinson, J.M., and White, L.D., 1992, The dynamic relationship between ground water and the Columbia River—Using deuterium and oxygen-18 as tracers: Journal of Hydrology, v. 135, p. 1–12.

McDonald, M.G., and Harbaugh, A.W., 1988, A modular three-dimensional finite-difference ground-water flow model: U.S. Geological Survey Techniques of Water-Resources Investigations, book 6, chap. A1, 586 p.

McGuire, V.L., Johnson, M.R., Schieffer, R.L., Stanton, J.S., Sebree, S.K., and Verstraeten, I.M., 2003, Water in storage and approaches to ground-water management, High Plains aquifer, 2000: U.S. Geological Survey Circular 1243, 51 p.

Miller, C.D., Durnford, Deanna, Halstead, M.R., Altenhofen, Jon, and Flory, Val, 2007, Stream depletion in alluvial valleys using the *SDF* semianalytical model: Ground Water, v. 45, no. 4, p. 506–514.

Morel-Seytoux, H.J., 1975, A simple case of conjunctive surface-ground-water management: Ground Water, v. 13, no. 6, p. 506–515.

Morel-Seytoux, H.J., and Daly, C.J., 1975, A discrete kernel generator for stream-aquifer studies: Water Resources Research, v. 11, no. 2, p. 253–260.

Morrissey, D.J., 1989, Estimation of the recharge area contributing water to a pumped well in a glacial-drift, river-valley aquifer: U.S. Geological Survey Water-Supply Paper 2338, 41 p.

Mueller, F.A., and Male, J.W., 1993, A management model for specification of groundwater withdrawal permits: Water Resources Research, v. 29, no. 5, p. 1359–1368.

Myers, N.C., Jian, Xiaodong, and Hargadine, G.D., 1996, Effects of pumping municipal wells at Junction City, Kansas, on streamflow in the Republican River, northeast Kansas, 1992–94: U.S. Geological Survey Water-Resources Investigations Report 96–4130, 58 p.

National Research Council, 2008, Prospects for managed underground storage of recoverable water: Washington, D.C., The National Academies Press, 337 p.

Newsom, J.M., and Wilson, J.L., 1988, Flow of ground water to a well near a stream—Effect of ambient ground-water flow direction: Ground Water, v. 26, no. 6, p. 703–711.

Nyholm, Thomas, Christensen, Steen, and Rasmussen, K.R., 2002, Flow depletion in a small stream caused by ground water abstraction from wells: Ground Water, v. 40, no. 4, p. 425–437.

Nyholm, Thomas, Rasmussen, K.R., and Christensen, Steen, 2003, Estimation of stream flow depletion and uncertainty from discharge measurements in a small alluvial stream: Journal of Hydrology, v. 274, p. 129–144.

Peralta, R.C., Kowalski, K., and Cantiller, R.R.A., 1988, Maximizing reliable crop production in a dynamic stream/aquifer system: Transactions of the American Society of Agricultural Engineers, v. 31, no. 6, p. 1729–1736.

Peterson, S.M., Stanton, J.S., Saunders, A.T., and Bradley, J.R., 2008, Simulation of ground-water flow and effects of ground-water irrigation on base flow in the Elkhorn and Loup River Basins, Nebraska: U.S. Geological Survey Scientific Investigations Report 2008–5143, 65 p.

Pollock, D.W., 1994, User's guide for MODPATH/MODPATH-PLOT, Version 3—A particle tracking post-processing package for MODFLOW, the U.S. Geological Survey finite-difference ground-water flow model: U.S. Geological Survey Open-File Report 94–464, 248 p.

Pool, D.R., and Dickinson, J.E., 2007, Ground-water flow model of the Sierra Vista Subwatershed and Sonoran portions of the Upper San Pedro Basin, southeastern Arizona, United States, and Northern Sonora, Mexico: U.S. Geological Survey Scientific Investigations Report 2006–5228, 48 p.

Prudic, D.E., Niswonger, R.G., and Plume, R.W., 2006, Trends in streamflow on the Humboldt River between Elko and Imlay, Nevada, 1950–99: U.S. Geological Survey Scientific Investigations Report 2005–5199, 58 p.

Reeves, H.W., 2008, STRMDEPL08—An extended version of STRMDEPL with additional analytical solutions to calculate streamflow depletion by nearby pumping wells: U.S. Geological Survey Open-File Report 2008–1166, 22 p.

Reeves, H.W., Hamilton, D.A., Seelbach, P.W., and Asher, A.J., 2009, Ground-water-withdrawal component of the Michigan water-withdrawal screening tool: U.S. Geological Survey Scientific Investigations Report 2009–5003, 36 p.

Reichard, E.G., 1995, Groundwater-surface water management with stochastic surface water supplies—A simulation optimization approach: Water Resources Research, v. 31, no. 11, p. 2845–2865.

Reilly, T.E., Franke, O.L., and Bennett, G.D., 1987, The principle of superposition and its application in ground-water hydraulics: U.S. Geological Survey Techniques of Water-Resources Investigations, book 3, chap. B6, 28 p.

Risley, J.C., Constantz, Jim, Essaid, Hedeff, and Rounds, Stewart, 2010, Effects of upstream dams versus ground-water pumping on stream temperature under varying climate conditions: Water Resources Research, v. 46, W06517, 32 p. (Also available at *http://dx.doi.org/10.1029/2009WR008587.*)

Rosenberry, D.O., and LaBaugh, J.W., eds., 2008, Field techniques for estimating water fluxes between surface water and ground water: U.S. Geological Survey Techniques and Methods 4–D2, 128 p.

Rutledge, A.T., 1998, Computer programs for describing the recession of ground-water discharge and for estimating mean ground-water recharge and discharge from streamflow records—Update: U.S. Geological Survey Water-Resources Investigations Report 98–4148, 43 p.

Sheets, R.A., Darner, R.A., and Whitteberry, B.L., 2002, Lag times of bank filtration at a well field, Cincinnati, Ohio, U.S.A.: Journal of Hydrology, v. 266, p. 162–174.

Singh, S.K., 2003, Flow depletion of semipervious streams due to pumping: Journal of Irrigation and Drainage Engineering, v. 129, no. 6, p. 449–453.

Sophocleous, M., 2000, From safe yield to sustainable development of water resources—The Kansas experience: Journal of Hydrology, v. 235, p. 27–43.

Sophocleous, Marios, 2012, On understanding and predicting groundwater response time: Ground Water, v. 50, no. 4, p. 528–540.

Sophocleous, Marios, Koussis, Antonis, Martin, J.L., and Perkins, S.P., 1995, Evaluation of simplified stream-aquifer depletion models for water rights administration: Ground Water, v. 33, no. 4, p. 579–588.

Sophocleous, Marios, Townsend, M.A., Vogler, L.D., McClain, T.J., Marks, E.T., and Coble, G.R., 1988, Experimental studies in stream-aquifer interaction along the Arkansas River in central Kansas—Field testing and analysis: Journal of Hydrology, v. 98, p. 249–273.

Spalding, C.P., and Khaleel, Raziuddin, 1991, An evaluation of analytical solutions to estimate drawdowns and stream depletions by wells: Water Resources Research, v. 27, no. 4, p. 597–609.

Stanton, J.S., Peterson, S.M., and Fienen, M.N., 2010, Simulation of groundwater flow and effects of groundwater irrigation on stream base flow in the Elkhorn and Loup River Basins, Nebraska, 1895–2055—Phase two: U.S. Geological Survey Scientific Investigations Report 2010–5149, 78 p.

Stark, J.R., Armstrong, D.S., and Zwilling, D.R., 1994, Stream-aquifer interactions in the Straight River area, Becker and Hubbard Counties, Minnesota: U.S. Geological Survey Water-Resources Investigations Report 94–4009, 83 p.

Stonestrom, D.A., and Constantz, Jim, 2003, Heat as a tool for studying the movement of ground water near streams: U.S. Geological Survey Circular 1260, 96 p.

Su, G.W., Jasperse, James, Seymour, Donald, Constantz, James, and Zhou, Quanlin, 2007, Analysis of pumping-induced unsaturated regions beneath a perennial river: Water Resources Research, v. 43, W08421, 14 p. (Also available at *http://dx.doi.org/10.1029/2006WR005389.*)

Theis, C.V., 1940, The source of water derived from wells—Essential factors controlling the response of an aquifer to development: Civil Engineering, v. 10, no. 5, p. 277–280.

Theis, C.V., 1941, The effect of a well on the flow of a nearby stream: Transactions of the American Geophysical Union, v. 22, no. 3, p. 734–738.

Turnipseed, D.P., and Sauer, V.B., 2010, Discharge measurements at gaging stations: U.S. Geological Survey Techniques and Methods, book 3, chap. A8, 87 p.

Wahl, K.L., and Tortorelli, R.L., 1997, Changes in flow in the Beaver–North Canadian River Basin upstream from Canton Lake, western Oklahoma: U.S. Geological Survey Water-Resources Investigations Report 96–4304, 56 p.

Wallace, R.B., Darama, Yakup, and Annable, M.D., 1990, Stream depletion by cyclic pumping of wells: Water Resources Research, v. 26, no. 6, p. 1263–1270.

Walton, W.C., 2010, Aquifer system response time and groundwater supply management: Ground Water, v. 49, no. 2, p. 126–127.

Wilson, J.L., 1993, Induced infiltration in aquifers with ambient flow: Water Resources Research, v. 29, no. 10, p. 3503–3512.

Winslow, J.D., 1962, Effect of stream infiltration on groundwater temperatures near Schenectady, New York: U.S. Geological Survey Professional Paper 450–C, p. 125–128.

Winter, T.C., Harvey, J.W., Franke, O.L., and Alley, W.M., 1998, Ground water and surface water—A single resource: U.S. Geological Survey Circular 1139, 79 p.

Young, R.A., and Bredehoeft, J.D., 1972, Digital computer simulation for solving management problems of conjunctive groundwater and surface water systems: Water Resources Research, v. 8, no. 3, p. 533–556.

Zarriello, P.J., and Ries, K.G., III, 2000, A precipitation-runoff model for analysis of the effects of water withdrawals on streamflow, Ipswich River Basin, Massachusetts: U.S. Geological Survey Water-Resources Investigations Report 00–4029, 99 p.

Zhang, Yingqi, Hubbard, Susan, and Finsterle, Stefan, 2011, Factors governing sustainable groundwater pumping near a river: Ground Water, v. 49, no. 3, p. 432–444.

Zheng, Chunmiao, and Wang, P.P., 1999, MT3DMS—A modular three-dimensional multispecies transport model for simulation of advection, dispersion and chemical reactions of contaminants in ground-water systems—Documentation and user's guide: U.S. Army Corps of Engineers Contract Report SERDP–99–1.

Zlotnik, V.A., 2004, A concept of maximum stream depletion rate for leaky aquifers in alluvial valleys: Water Resources Research, v. 40, W06507, 9 p. (Also available at *http://dx.doi.org/10.1029/2003WR002932.*)

Zlotnik, V.A., and Tartakovsky, D.M., 2008, Stream depletion by groundwater pumping in leaky aquifers: Journal of Hydrologic Engineering, v. 13, no. 2, p. 43–50.